LIQUID CRYSTALS:
THEIR PHYSICS, CHEMISTRY
AND APPLICATIONS

LIQUID CRYSTALS:
THEIR PHYSICS,
CHEMISTRY
AND APPLICATIONS

PROCEEDINGS OF
A ROYAL SOCIETY DISCUSSION MEETING
HELD ON 27 AND 28 OCTOBER 1982

ORGANIZED AND EDITED BY
C. HILSUM, F.R.S., AND E. P. RAYNES

LONDON
THE ROYAL SOCIETY
1983

Printed in Great Britain for the Royal Society
at the
University Press, Cambridge

ISBN 0 85403 210 X

First published in *Philosophical Transactions of the Royal Society of London*,
series A, volume 309 (no. 1507), pages 69–239

Copyright

Published by the Royal Society
6 Carlton House Terrace, London SW1Y 5AG

PREFACE

Modern electronics requires special types of display devices. The low voltage and power budget is incompatible with the standard cathode-ray tube, which is in any case too large. The invention of the liquid crystal display was welcomed as the most promising answer to the problem and today, after a few setbacks and disappointments, this first hope is being fulfulled. Liquid crystal watch and calculator displays are seen in most homes, and the more taxing applications as graphical and picture displays are the topic of research in many countries.

Liquid crystals are complex materials. Although macroscopic theories frequently allow accurate descriptions in terms of empirical constants, the microscopic theories usually reflect the inherent complexity of the liquid crystalline state. The typical mixture used in a practical display is a multi-component eutectic, and the research to devise and synthesize the individual components, and to predict the chemical, physical and electro-optic properties of the mixture, provides challenging tasks for workers in a number of disciplines. In a field of such intense commercial rivalry, it is natural that progress will be both rapid and shrouded. The Discussion Meeting held by the Royal Society in October 1982 provided a venue for world authorities to present their views on current progress, and to chart the promising paths for the future.

As organizers of the meeting, we attempted a broad coverage, inviting speakers who could review most aspects of the physics, chemistry, and applications of liquid crystals. Fifteen of their papers have been collected in this book. It gives a unique opportunity for the expert and the untutored to survey a topic of great breadth and depth.

C. Hilsum

E. P. Raynes

April 1983

CONTENTS

† Elected F.R.S. 17 March 1983.

Phil. Trans. R. Soc. Lond. A **309**, 71–75 (1983)
Printed in Great Britain

Introductory remarks

BY F. C. FRANK, F.R.S.

H. H. Wills Physics Laboratory, The University, Royal Fort,
Tyndall Avenue, Bristol BS9 1TL, U.K.

Liquid crystals have been recognized as such for just about 100 years. The pure substances known to exhibit one or more liquid crystal phases are numbered in thousands: Professor Gray will probably update us on the present score. The liquid crystals give us a very fine object lesson in the interaction of science and application. Their general nature has been well understood since Georges Friedel's great review article in 1921. There was not much written about them in English before the Faraday Society discussion of 1933 – the meeting at which Sir William Bragg brilliantly overnight perceived the geometrical origin of the focal conic texture of smectics, unaware that Friedel had explained it all a dozen years before. After that, most scientists closed the book. The liquid crystals were scientific curiosities, basically understood, and no one had a use for them. Chemists gradually discovered more mesomorphic substances but, on the whole, knowledge of their physics went backwards rather than forwards. The history of the science of liquid crystals is marked with repeated rediscovery of things forgotten. When, in 1958, I published a paper about liquid crystals I was advocating renewed study of them as a source of information about molecular interactions. However, for one thing, that was not too easy, except in terms of severely simplified simulation models; for another, scientific curiosity alone was not sufficient incentive.

In the 1960s there was a fresh appreciation of the need for new methods of display, to make electrically generated signals visible to the eye. Solid state electronics had brought it about that all of a complex electronic system was minute in bulk, weight and power consumption in comparison with its associated display. At last, the science of liquid crystals acquired a sufficient number of target points to engage the attention of the several dozen new research workers needed to advance knowledge and understanding significantly beyond the level attained by the first half-dozen pioneers, and to a considerable extent forgotten in the interim.

The best of the light-valve phenomena exploited so far is that of the electrically switched twisted nematic. The entertaining optical properties of the twisted nematic were elucidated both experimentally and theoretically by Mauguin in 1911. Mauguin's results were to be rediscovered, more than once, about 60 years later, generally by less elegant treatments and without mention of Mauguin. He did not study the electrical switching of the twisted nematic, but with hindsight that was obvious after the work of Freedericksz in 1927. The invention latent in the combination of these two known phenomena lay dormant till the need for application called it forth in 1971.

This is far from being the only useful application of the science of liquid crystals, and we should hear of others in the course of the Discussion but this is the one that has been responsible for the renascence of the science.

On terminology

I believe one of the most valuable things I can do by way of introduction to this discussion is to make some comments on the terminology of the subject, which is indeed somewhat confused and confusing. I shall speak dogmatically. You need not agree with me; but I think I may reasonably ask that where what I say gives some indication of possible misunderstanding you take care to define your terms. I first take the term *liquid crystal* and illustrate it by an example of what is *not* a liquid crystal. The *colloid crystal* you obtain with a dilute suspension of negatively charged silica or polystyrene spheres in water at minimal ionic strength is a *crystal*. It is an extreme example of the *plastic crystal*. The particles array themselves on a body-centred cubic lattice that can be seen directly with the microscope or inferred from the coloured Bragg reflexions in visible light. That crystal is liquid in common parlance. It can be poured from one test-tube to another. But it is not liquid: its yield stress is very small, but finite. So it is not a liquid crystal: nor is it a mesomorph. It is a *crystal* belonging to the same structural class as body-centred-cubic iron. The tallest self-supporting uniform column of iron, if the wind didn't blow, would be of the order 10 km in height: the corresponding height for the colloid crystal is of the order of a millimetre, and it is only as large as that because of the buoyant support of the water. A factor of 10^{10} is just a large number, and there is still all the difference in the world between a very large number and infinity. A liquid has strictly zero yield stress. Anything having a three-dimensional space-lattice, with its lattice points occupied by material objects, will not have zero yield stress. Dislocations in that lattice structure may lower the yield stress by a factor of 10^3, say; but 10^{13} is just as far short of infinity as 10^{10}. There is inevitably some anchoring force on the dislocations. If there were none, the dislocations would escape.

Why did I take care to specify that the lattice points were occupied by material objects? I shall answer that question presently.

Nematic, cholesteric, smectic, the three Friedelian classes: etymology gives us little assistance in interpreting those names. 'Nematic' means 'thready', but disclinations, which can look like threads, are not solely confined to nematic phases. 'Cholesteric' derives from the historical fact that the first recognized, the first liquid crystals recognized in fact, were derivatives of cholesterol, but most cholesterics are not. 'Smectic' means 'soapy', but not everything soapy is smectic. Fortunately the first two at least have acquired stable and unquestioned meanings: *nematic*, the simple uniaxial liquid, and *cholesteric* its modification, spontaneously twisting about a single transverse axis, induced by optical activity in the molecules or optically active additives.

'*Smectic*' is the one that gives trouble. Friedel recognized only one smectic structure, layered, with no long-range positional correlation within the layers. We have since learnt that specification does not uniquely define a structure or a phase, and one substance can have more than one distinct phase satisfying that specification. Hence the necessity to subclassify the smectics; but I would allow none to be called smectics of any kind that do not satisfy that specification. And I believe that all those thus misnamed are not even liquid crystals, they are plastic crystals. I would tolerate, and even advocate, a word such as 'smectoid' to recognize their relation to the smectics.

Thermotropic and lyotropic: that is a false contrast; but though the naming is illogical, it serves to make a significant and valuable distinction. It is well established and we should tolerate it.

Blue phases: I spoke of the cholesteric twisting about a single axis. It is not topologically possible to twist uniformly through many turns about more than one axis: but twisting about two axes at once over a short distance is possible, and there are substances that achieve that condition, making a three-dimensional lattice of twistedness with overall cubic symmetry. They occur in a half-degree temperature range between cholesteric and isotropic in substances giving cholesterics of short pitch, say 300 nm. They are called blue phases. The blue colour is Bragg reflexion from the lattice of twistedness, the lattice parameter of which is of the same order of magnitude as the pitch of the related cholesteric. There are blue phases of other colour than blue – that is entirely in the spirit of liquid crystal etymology. In that $\frac{1}{2}$K temperature range there can be two different cubic phases, and an amorphous one, picturesquely called the *blue fog*, in which the lattice of twistedness melts.

Now, apart from the last of these, we have three-dimensional lattices; so can these materials be liquid, with zero yield stress? I think they can. The lattice points are not occupied by distinct material objects but by states of orientation. I think matter can pass through the lattice points, the body plastically deforming while the lattice does not. So they *are* liquid crystals. What about the amorphous blue phase? Must it be excluded from the liquid crystal family because it is not a crystal? I think not, since locally it is still effectively nematic.

These blue phases make a thermodynamically and structurally distinct class. They were known before Friedel's time, though not structurally interpreted until much later. I think they should be recognized as a fourth class, after the three Friedelian classes of liquid crystals.

I think they may not be mentioned again in this discussion. They are totally useless, I think, except for one important intellectual use, that of providing tangible examples of topological oddities, and so helping to bring topology into the public domain of science, from being the private preserve of a few abstract mathematicians and particle theorists.

Discotic and columnar: it is not too late to insist on a little rationality here. 'Discotic' is a word sensibly describing the shape of a molecule rather than a phase. *Discotic nematics* do exist. That was what Chandrasekhar was looking for when instead he discovered the first non-lyotropic columnar phase. His disc-shaped molecules assembled themselves in columns, with no long-range positional correlation along their length, packed parallel in a two-dimensional lattice. That is a distinct structural class, for which I think the best name is *columnar*. Counting the blue phases as the fourth general class, this is the fifth. Like the smectic class, it has subdivisions according to the nature of its two-dimensional lattice.

Biaxial nematics: I do not know whether we shall be told anything about biaxial nematics in this Discussion. In case we are, let me remark that the term 'biaxial' belongs to an *optical* classification of symmetries. In crystal optics, homogeneous bodies are biaxial, uniaxial or isotropic. In classifying the liquid crystals, however, we are concerned basically with structural symmetry, which offers more varieties than that. An optically isotropic body may be structurally isotropic, for example glass, or it may be a cubic crystal. An optically uniaxial body may be simply uniaxial, for example the ordinary nematic, or it may be a hexagonal or tetragonal crystal. An optically biaxial crystal may be structurally orthorhombic, monoclinic or triclinic; the two latter may be improbable for nematic symmetries, but I do not know that they are impossible. They will impose different constraints on the possible defects in the structure. When de Gennes speaks of a biaxial nematic I understand him to mean a structure having no long-range positional correlation in any direction, with orthorhombic symmetry for its orientational distribution functions. If so, I think it would be better called *orthorhombic nematic*.

It is a matter for judgement whether this should be considered a sixth class of liquid crystals. I prefer to regard it as a subclass of the nematics, and stick at five.

[3]

Homeotropic, homogeneous: 'homeotropic', to describe either a boundary condition, or a condition throughout a layer between plane parallel walls, in which the director is perpendicular to the wall or walls, is a useful long-established word, with no alternative meaning. 'Homogeneous' to describe the infinite set of converse conditions with the director parallel to the wall, but not necessarily uniformly oriented, is ridiculous. 'Homogeneous' has an ordinary dictionary meaning, for which 'uniform' is a synonym (in French or German, as well as in English): it is unforgivable to give a technical meaning to a word in conflict with its ordinary meaning when the ordinary meaning has applications in the same context. It would be good to have a different technical term: 'planar', 'parallel' and 'horizontal' are each in various ways objectionable but all vastly preferable to 'homogeneous', which should be totally avoided, saying 'uniform' when in any liquid crystal context the ordinary meaning of 'homogeneous' is intended.

Direction of the molecules, molecular long axis, director: I shall be surprised if someone in this Discussion does not make reference to the *direction of the molecules*, or of their *long axes*. Let me remind you that these are concepts having precise meaning only for a simplified model of the liquid crystal, in which the molecules are axisymmetric bodies, which they never really are. These concepts are incapable of precise definition for real molecules. The *director* is a good concept, at least for the nematics. It is the local symmetry axis for the distribution functions of any orientational property of the molecules, or of intermolecular separations. The mean direction of molecular long axes may coincide with the director, but in discotics it may be the direction of their short axes instead. In smectics, a symmetry axis for orientational distribution functions can exist if, and only if, it coincides with the normal to the layers. *Inclination of the molecules* in smectic C has no exact definition except for the simplified models.

Order parameter is in general a term well defined only for simplified simulacra of reality. It is true that in phase-transition theory one order parameter can dominate and determine the nature of a transition, but it is an obvious fallacy to suppose that in general the character of molecular order or disorder is simple enough to be described and measured by a single parameter. The order parameters determined by optical, magnetic or n.m.r. measurements are equivalent only for the simplified models. If they differ by 10%, that reflects reality.

The very word *molecule* can itself be a trap. In mesomorphic polymers it must mean only the whole polymeric long chain, which must not be confused with the orientable entities that give it mesomorphic character.

Enough. I hope I do not sound like a voice crying in the wilderness, but that what I have said does a little to make straight the path of those who are to follow.

Discussion

M. G. CLARK (*R.S.R.E., Malvern, Worcs., U.K.*). I agree that there are difficulties in defining the 'long axis' if only static properties of the nematic are considered. However, there is a natural and unambiguous definition of the long axis if one considers the dynamical theory. The properties of practical interest correspond to timescales longer than those for which molecular inertia is important, with static properties as the long timescale limit. Thus the long axis may be defined as the largest-eigenvalue principal axis of the rotational diffusion tensor

$$\mathbf{D}_{\mathrm{rot}} = \int_0^\infty \mathrm{d}t \langle \boldsymbol{\omega}(0)\,\boldsymbol{\omega}(t)\rangle,$$

[4]

where ω is the molecular angular velocity. This means that the 'long axis' is the axis about which rotational diffusion is fastest, which is the natural definition for both elongated and disc-shaped molecules. The definition does not assume any particular molecular symmetry. From the definition we see that the long axis is defined in the molecular frame by the local dynamical structure of the particular fluid, and that its orientation relative to the molecular geometry can and should be determined experimentally. (See M. F. Bone, A. H. Price, M. G. Clark & D. G. McDonnell, in *Liquid crystals and ordered fluids* (ed. J. F. Johnson & A. C. Griffin), vol. 4, Plenum, New York (in the press).)

On the question of agreement between different methods of measuring nematic order parameter, it should be noted that Tough & Bradshaw (in preparation) have been shown that when a well conditioned extrapolation procedure is used, the curves of order parameter against temperature derived from magnetic susceptibility data and refractive index data are in very close agreement. This was shown for both the materials **1** and **2**, which are far from symmetrical.

F. C. FRANK. I cannot agree with Dr Clark. At best it seems to me that the axis of fastest rotational diffusion is just one more of the directional properties that can be associated with a molecule plus its environment, but I would say it should on no account be called the long axis when it may, as he indicates, alternatively be a short axis of the molecule in the ordinary geometrical sense of short and long.

Phil. Trans. R. Soc. Lond. A **309**, 77–92 (1983)
Printed in Great Britain

The chemistry of liquid crystals

By G. W. Gray

Department of Chemistry, University of Hull, Hull HU6 7RX, U.K.

The nature of the liquid crystal phases formed by different compounds and the thermal and other physical characteristics of these phases are strongly affected by the stereochemistry and structure of the molecules. Precise structure–property correlations of an embracing nature cannot yet be quantified, but even a qualitative understanding of such relations is desirable from many standpoints, including that of the achievement of still better liquid crystal materials for the ever widening range of applications that is emerging for these ordered but fluid systems. By means of a range of selected examples, an attempt is made to illustrate not only those areas where generalizations seem possible, but also others wherein the properties of the materials are much more difficult to understand, to such an extent that doubt must be cast on at least some theoretical concepts of the liquid crystal state.

Introduction

This paper is concerned with the way in which particular features of molecular structure have been found to influence the liquid crystal properties – the type(s) of liquid crystal phase and the liquid crystal transition temperatures – of a system. Considering liquid crystals in their broadest sense, this is a vast subject and it is necessary to begin by narrowing down our objectives, as shown in figure 1.

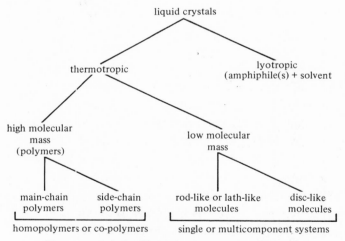

Figure 1. Subclassifications of liquid crystal systems.

With regard to figure 1, the following points should be made.

1. Lyotropic liquid crystal systems composed of amphiphilic compounds and solvent will not be considered further, and attention will be concentrated upon thermotropic systems that form their liquid crystal phases as a result of temperature change: heating or cooling.

2. Thermotropic liquid crystal systems may be of either high molecular mass (polymer

systems) or low molecular mass. Because the former are the subject of a subsequent presentation at this meeting, I shall deal here only with low molecular mass systems.

3. Low molecular mass liquid crystal systems may themselves be divided into those involving elongated, rod-like molecules and those involving disc-shaped molecules. It is my intention to deal only with the rod-like molecular systems, for the discotic systems are the subject of Professor Chandrasekhar's contribution to this symposium.

My theme will be to stress how very sensitive liquid crystal properties are to small perturbations of molecular structure, with the result that sweeping generalizations are simply not possible. The situation is certainly not quantifiable. We are in fact concerned with how well rod-shaped molecules are disposed to fit together into a smectic or nematic liquid crystal lattice, and our ability to comprehend exactly how molecular structure, shape, polarizability and flexibility influence such three-dimensional packing is, to say the least, limited.

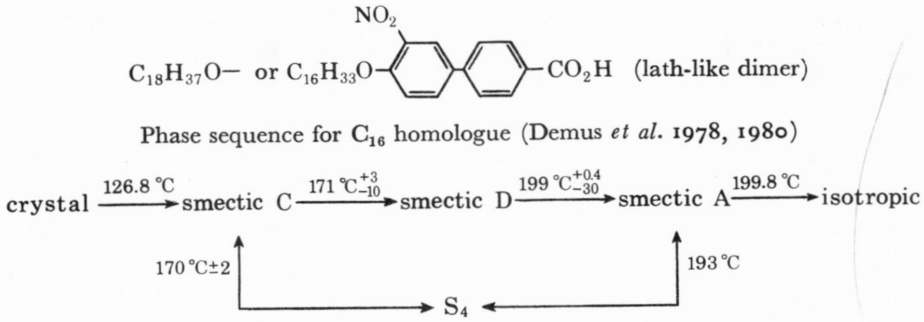

$$C_{18}H_{37}O- \text{ or } C_{16}H_{33}O-\underset{NO_2}{\diagup}-CO_2H \quad \text{(lath-like dimer)}$$

Phase sequence for C_{16} homologue (Demus *et al.* 1978, 1980)

crystal $\xrightarrow{126.8\,°C}$ smectic C $\xrightarrow{171\,°C_{-10}^{+3}}$ smectic D $\xrightarrow{199\,°C_{-30}^{+0.4}}$ smectic A $\xrightarrow{199.8\,°C}$ isotropic

$170\,°C\pm2$ $\qquad\qquad\qquad\qquad\qquad\qquad 193\,°C$

$\longrightarrow S_4 \longleftarrow$

S_D, optically isotropic, viscous, cubic;
S_4, birefringent (discotic?) (Lydon 1981).
FIGURE 2. The phase behaviour of some smectic D materials.

Consider for example the rod-like molecules of the dimeric 4'-alkoxy-3'-nitrobiphenyl-4-carboxylic acids shown in figure 2. The elongated dimeric molecules appear to have exactly the right structure to form smectic or nematic thermotropic phases of the classical kind involving statistically parallel alignments of the rods. The C_{16} and C_{18} homologues do indeed form such phases, but injected between their ordinary S_C and S_A or isotropic phases lies an optically isotropic cubic phase or, alternatively, depending upon the experimental conditions relating to the heating–cooling cycles, a different phase (S_4), which could possibly be discotic in nature. Lydon (1981) has rather neatly tried to rationalize this apparently contradictory situation by proposing that the elongated dimeric molecules can, under suitable conditions of temperature, form disc-like aggregates that may then stack up to form columns. These columns may then either arrange themselves parallel to give a columnar discotic phase or form jointed rod structures, two of which, as pointed out by Luzzati & Spegt (1967) and Tardieu & Luzzati (1970), may interconnect to form a structure with overall cubic symmetry – a structure that would be compatible with the properties of the S_D phase (Tardieu & Billard 1976). This illustrates how judgements of phase behaviour based on an over-simplistic view of molecular structure and packing can be very misleading.

The unexpected discotic behaviour of diisobutylsilanediol (Bunning *et al.* 1980, 1982) provides another interesting example of this kind.

[8]

The above examples relate to major effects arising from the manner in which molecules pack together. If we consider that subtle features of molecular shape and packing may influence the subtleties of the phase behaviour of liquid crystals, the magnitude of the problem of establishing exact structure–phase correlations for thermotropic liquid crystal systems is realized.

ROD-LIKE MOLECULAR STRUCTURES

The importance of an elongated, rod-like molecular shape in relation to liquid crystal phases was clearly established early in the century. Some typical compounds that exhibit thermotropic liquid crystal phases are shown in figure 3.

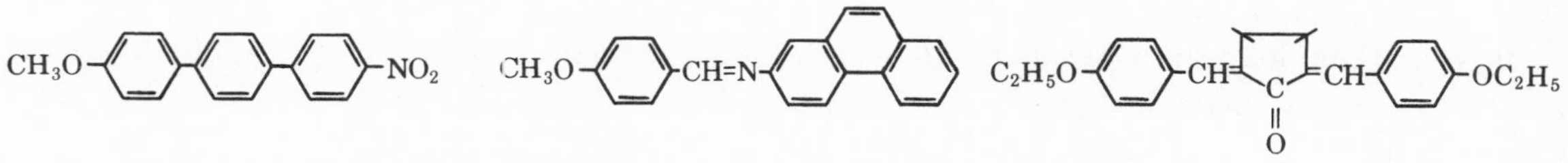

FIGURE 3. The structures of some typical compounds (nematogens) that form nematic phases.

The structural relation is very clear, for as can be seen from the formulae in figure 3, the molecules contain rings and double bonds, which help to prevent the molecules from adopting nonlinear conformations or configurations that would militate against the parallel packing of rods needed for formation of liquid crystal states. The concept of a rod-like molecular structure as a necessary requirement for the formation of liquid crystal phases has therefore become generally accepted and applies to smectic, nematic and cholesteric phases.

(a) *Smectic phases*. With the exception of the S_D phase, smectic phases are lamellar in nature. Smectic phases (in the plural) is written because we must remember the complexities that arise as a result of the polymorphism of smectic systems. Many structural variants of the smectic phase are known, dependent upon how the rod-like molecules are arranged in the smectic lamellae, the tilt angle of the rods with respect to the lamellar planes, and the degree of correlation of structure from layer to layer. Table 1 summarizes the state of current knowledge, showing the division of smectic systems into (1) liquid crystal smectic phases with no or low correlation of structure between layers, and (2) crystal type smectic phases with extensive interlayer correlations. Subdivisions within these categories depend upon molecular tilt and packing within the layers.

TABLE 1. SMECTIC POLYMORPHIC TYPES

liquid crystal smectics			crystal-type smectics		
no or low correlation of order between layers			long-range correlation of order between layers		
type	lamellar order	tilt	type	lamellar order	tilt
S_A	none	no	S_B	yes	no
S_B (hexatic)	yes	no	S_E	yes	no
S_C	none	yes	S_G; $S_{G'}$	yes	yes
S_F	yes	yes	S_H; $S_{H'}$	yes	yes
S_I	yes	yes			

(b) *Nematic phases*. Here we are involved with a simple non-lamellar, statistical parallel packing of rods.

(c) *Cholesteric phases*. These phases are simply spontaneously twisted nematic phases, the twist arising from the optically active nature of the chiral molecules, or alternatively from the chirality of suitable solutes dissolved in a nematic host. Cholesteric phases are therefore formed by nematogenic molecular systems that are optically active. Therefore from a structural standpoint, nematogenic and cholesterogenic compounds can be considered together.

THE RELATIVE SMECTIC–NEMATIC TENDENCIES OF A COMPOUND

Even if we ignore the problems of smectic polymorphism, it is not easy to make more than general comments about the relative smectic–nematic tendencies that a compound may possess. The question is apparently a most basic one, but it is in fact very subtle, for we are attempting to judge features of molecular structure that influence not simply whether the molecules will pack with their long axes parallel, but whether they will do so in a manner such that their ends lie in planes, as distinct from an interdigitated nematic arrangement. This is clearly not a simple matter.

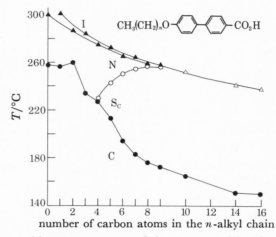

FIGURE 4. The transition temperatures and the sequence of liquid crystal phases exhibited as a function of chain length n, for a homologous series of carboxylic acids.

However, it did seem clear – at least until recently – that lengthening the alkyl chain of a terminal n-alkyl or n-alkoxy substituent did strongly promote smectic character through a mutual affinity of neighbouring alkyl chains. Numerous examples of this are known, and one showing the phase distribution in a homologous series is shown in figure 4. However, even this generalization is not completely true. Consider the bicyclohexyl compounds – members of the CCH series (Pohl *et al.* 1978) – of the structure shown in figure 5. For this series, smectic properties *decrease* as R increases in length, and as explained by Brownsey & Leadbetter (1981) a particular packing situation has been favoured by a propitious combination of the best anti-parallel correlation of the cyano groups for this alicyclic system and the similar sizes of the terminal groups. The behaviour can be comprehended, but it is doubtful that it could have been predicted.

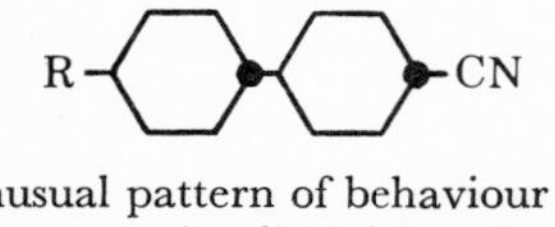

FIGURE 5. The general structure and an unusual pattern of behaviour for the bicyclohexyl compounds known as the CCH materials. Smectic properties diminish as R = alkyl increases in length.

Smectic systems

The question now arises as to molecular structural features that may favour particular smectic types from the range of possible smectic polymorphic modifications. As yet no clear relations are obvious, perhaps not surprisingly because we are now attempting to relate molecular structure to detailed aspects of particular lamellar packings. We are in fact almost attempting to predict solid crystal structure from molecular structure, and we are all well aware that even the ascent of a homologous series of alkyl-substituted compounds can cause dramatic changes in *crystal* lattice type. Similarly the ascent of a homologous series of mesogens can cause most marked fluctuations in the types of smectic phase formed. This is therefore a very difficult area on which to comment at present, when even the addition or removal of one CH_2 group in a terminal chain can dramatically alter the smectic phase type(s) exhibited.

Only in smectic C phases have clear attempts been made to relate molecular structure to the tendency of the compound to give the tilted, disordered lamellar structure that this phase adopts. In line with structural studies made by de Jeu (1977), there is no doubt that terminal alkoxy groups in aromatic mesogens favour formation of tilted S_C phases rather than orthogonal S_A phases. Because alkyl analogues of such materials are often S_A in character, models relating the tendency of the molecules to tilt (with respect to the lamellar planes) to the existence in the molecules of terminal outboard dipoles associated with the alkoxy substituents have been proposed (McMillan 1973); the molecules may or may not also carry a lateral central dipole. However, it is known that certain esters (Goodby *et al.* 1977) containing no significant terminal dipoles and only a central dipole do form S_C phases. These S_C phases do admittedly persist until lower temperatures than those of the analogous esters with terminal alkoxy substituents, so there is no doubt that terminal outboard dipoles do *enhance* S_C character, but they do not appear to be *essential* to the formation of the phase. Allowance for all these observations and their consequences in relation to McMillan's theory of S_C phases have been made by Van der Meer & Vertogen (1979).

Nematic–cholesteric systems

Until comparatively recently, most nematic–cholesteric materials studied in relation to the changes in phase transition temperatures brought about by change in molecular structure were aromatic in type, and these polarizable, rigid rings became regarded – wrongly as it emerges – as rather important to the formation of thermally stable liquid crystal phases. Through the study of such aromatic systems, certain ground rules were, however, established. These (Gray 1976) may perhaps usefully be summarized briefly here.

(*a*) *Some established structural relations for rod-shaped aromatic molecules*

(i) *Extension of the length of the rod-shaped molecule*

(1) *Addition of rings or multiple-bonded units to extend the rigid core of the molecule.* As can be seen from the examples given in figure 6, such extensions of the core structure result in significant increases in the nematic–isotropic liquid transition temperatures (T_{NI}).

(2) *Lengthening the terminal chain of an n-alkyl or n-alkoxy substituent.* The important point here is that although the effect on T_{NI} as the series is ascended varies from system to system (Gray 1962, 1976), the effects within a given homologous series are always regular. When the T_{NI} values are quite high, two falling T_{NI} curves may be drawn and reflect a regular, diminishing

alternation between odd and even members of the series. This common pattern of behaviour is evident in figure 4. This pattern changes, however, for systems exhibiting lower T_{NI} values. For example, for the well known series of PCH compounds (Eidenschink *et al.* 1977), the curves now rise, but the alternation effect is similar. Other systems can, however, show a combination of these two types, and this is found for the 4-*n*-alkyl-4'-cyanobiphenyls (Gray 1978), for which the initially falling T_{NI} values reach a minimum and then rise again. The alternation effect is again pronounced.

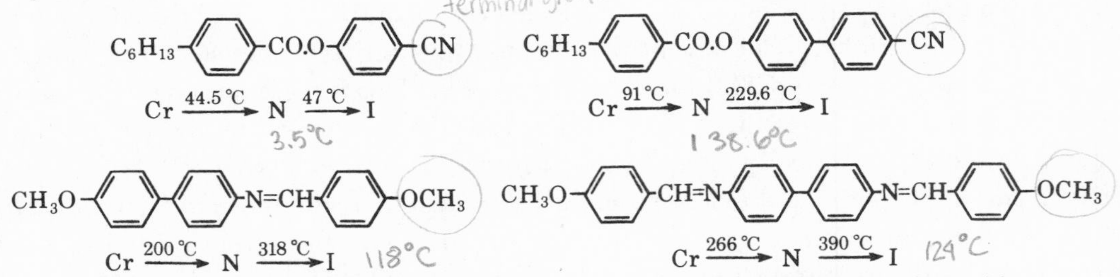

FIGURE 6. Data illustrating the enhancement of the nematic–isotropic liquid transition temperature (T_{NI}) associated with extending the core structure of a mesogen.

These results are not simple to interpret, and require (de Jeu *et al.* 1973) a combination of effects: the influence of the chain (odd or even) upon $\Delta\alpha$ (the anisotropy of the molecular polarizability) and the effect of temperature upon the molecular length : breadth ratio through its influence in causing deviations of the chain from an ideal all-*trans*, zig-zag conformation.

It is noted that a very striking effect arises (Gray & Harrison 1971 *a, b*) if the chain carries a bulky group such as phenyl at its end. Extremely large alternation effects (100–200°) occur, apparently as the ring moves on-axis and then off-axis for odd and even homologues.

(ii) *The effect of different terminal groups on* T_{NI}

A nematic terminal group efficiency order has been established for aromatic systems by their effects on T_{NI}: it is Ph > NHCOCH$_3$ > CN > OCH$_3$ > NO$_2$ > Cl > Br > N(CH$_3$)$_2$ > CH$_3$ > F > H. The high position occupied in the order by a terminal cyano group is noted, as is the higher position of alkoxy relative to alkyl. Any group is apparently superior to hydrogen in promoting T_{NI}. A different order exists for smectic phases.

(iii) *The effect of different linking groups in the core structure*

Figure 7 summarizes the order established for a number of different linking groups when these are used in molecules of the structure indicated. It is noted that a direct bond between the two benzene rings is the least effective linking unit.

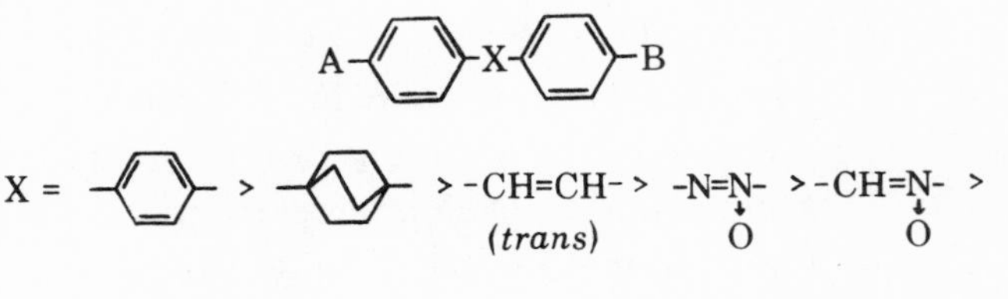

FIGURE 7. The relative effects of different linking groups between the aromatic rings in the core of the mesogen upon the nematic–isotropic liquid transition temperature (T_{NI}) (in order of decreasing T_{NI}).

[12]

(iv) The effects of different lateral substituents in the core structure

These effects were established by studies of acids of the structure shown in figure 8 (top).

The results are entirely consistent with maintaining a high molecular length:breadth ratio. The depressing effect of a lateral substituent (X) upon T_{NI} is proportionately related to the size of the substituent, and the permanent dipole of the ring–X bond is *not* a contributing factor. The T_{NI} depression is greatest for a given X if the molecule is short.

length: breadth ratio   & depressing effect

(1) *substituent not shielded*

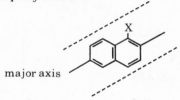

X ↙ SUBSTITUENT

T_{NI} falls in order of increasing substituent size (irrespective of permanent C–X dipole):
X = H > F > CH₃ ≈ Cl > Br > I ≈ NO₂.
Magnitude of decrease rises as molecular length decreases.

T_{NI} falls in order of increasing substituent size (irrespective of permanent C–X dipole): $X = H > F > CH_3 \approx Cl > Br > I \approx NO_2$. Magnitude of decrease rises as molecular length decreases.

WHAT IS MEANT BY "PARTIALLY SHIELDED"?

(2) *substituent partly shielded*

major axis

T_{NI} order: $X = Cl > Br > H > I$

FIGURE 8. A summary of the effects of different lateral substituents in the core of a mesogen upon the nematic–isotropic liquid transition temperature (T_{NI}).

TABLE 2. ENHANCED EFFECTS OF LATERAL SUBSTITUTION

(extra) molecular twisting (steric effect)

–N=CH– –OC₇H₁₅

X	$T_{NI}/°C$	$\Delta T_{NI}(\text{H} \to \text{Cl})/°C$	
H	163.5		
3Cl	96	67.5	
2′Cl	52	111.5	} enhanced twist
2Cl	45	118.5	
3′Cl	smectic	—	

As would now be expected (figure 8, lower part), a lateral substituent that for some structural reason does not broaden a molecule, as in suitable derivatives of naphthalene, can enhance T_{NI}. This is consistent with the length:breadth ratio's being maintained, but with a more efficient occupancy of space between the parallel-packed rod-shaped molecules.

Special cases exist for systems in which the lateral substituent not only broadens the molecule but also exerts a twisting steric effect, or as in the Schiff bases shown in table 2, an extra

[13]

twisting effect (the Schiff base linkage is itself not planar). The resulting increase in molecular thickness depresses T_{NI} very strongly. The much greater effect of the sterically twisting 2- and 2′-chloro substituents, relative to that of the 3-chloro substituent, is obvious.

(v) *The effects of branching of a terminal alkyl chain*

The effects are again well established (Gray & Harrison 1971 *a, b*) for systems such as those shown in table 3. The branching methyl group diminishes in its depressing effect upon T_{NI} on being moved towards the free end of the chain, but the very large effect of the 1-methyl substituent implies that a steric effect may operate here. Because cholesteric compounds are often produced (Gray & McDonnell 1975) by using chiral alkyl groups, which are necessarily branched, such effects are clearly important to the molecular engineering of cholesterogens.

TABLE 3. CHAIN BRANCHING

NC—⟨ ⟩—CH=N—⟨ ⟩—CH=CH-CO.Oalkyl

alkyl	$T_{NI}/°C$	comment
—CH₂CH₂CH₂CH₂CH₃	136.5	
—CHCH₂CH₂CH₂CH₃ (CH₃)	< 20	large effect
—CH₂CHCH₂CH₂CH₃ (CH₃)	112	
—CH₂CH₂CHCH₂CH₃ (CH₃)	103	gradual recovery
—CH₂CH₂CH₂CHCH₃ (CH₃)	119.5	

These, then, are some general effects established over the years for liquid crystal systems composed of aromatic molecules. Simple as some of the conclusions are, these rules have been applied usefully and they contributed to the discovery of the 4-alkyl- and 4-alkoxy-4′-cyanobiphenyls (Gray *et al.* 1973), which have been fundamental to the development of a liquid crystal display industry producing electro-optic displays of good quality and high lifetime capability. Table 4 shows how the rules were used. Central groups (A) such as —CH=N—, azo and azoxy were known to confer both colour and chemical–photochemical instability upon the compounds. Such groups were eliminated by directly linking the two rings, i.e. by a single bond. The depressing effect upon T_{NI} was countered by using for Y a group high in the nematic efficiency order, i.e. a cyano group (also needed for the strong positive dielectric anisotropy ($\Delta\epsilon$) needed for twisted nematic displays). The group X was chosen to be *n*-alkyl or *n*-alkoxy, and the resulting model systems were as formulated at the foot of table 4, and in several cases they represented the structures of compounds giving stable, colourless nematic phases at room temperature.

The interest aroused in these systems has resulted over the years in their very intensive study, and it has emerged that the high position of the cyano group in the T_{NI} efficiency order arises from the antiparallel association of the molecules. This was first demonstrated by Leadbetter *et al.* (1975), and occurs to minimize dipole–dipole repulsions between cyano groups. The situation for the biphenyls is shown in figure 9. For other cyano-substituted mesogens, the core overlap is only partial (PCH materials), and as already discussed, in the CCH systems, only the two atoms of the cyano groups are involved in the antiparallel overlap.

TABLE 4. DESIGN FOR A LOW-MELTING STABLE NEMATOGEN

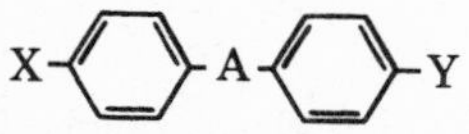

$$X-\bigcirc-A-\bigcirc-Y$$

Y counters X

(1) For maximum stability, A = single bond.
Consequence: very low T_{NI}.

(2) Promote T_{NI} by using a group Y high in nematic terminal group efficiency order,

e.g. Y = CN.

(3) Use X = alkyl or alkoxy to promote nematic order and maintain low melting point; this also gives a selection of homologues.

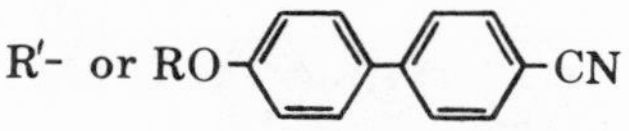

$$R'- \text{ or } RO-\bigcirc-\bigcirc-CN$$

$$R' = C_5H_{11}: \quad Cr \xrightarrow{22\,°C} N \xrightarrow{35\,°C} I$$

The thrust that the availability of the biphenyls gave to the display industry has naturally provided an incentive to produce alternative materials that might have modified properties that could be advantageous for other display forms. Of greatest significance in this respect has been the experimentation involving replacement of benzene rings by alicyclic rings.

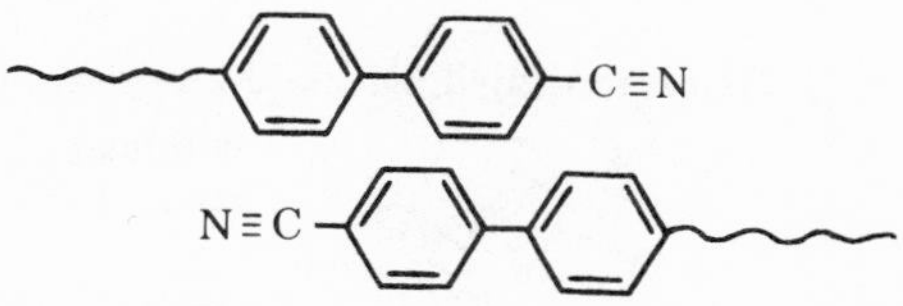

(180° rotation)

FIGURE 9. Antiparallel pairing or correlation of two molecules of a 4-alkyl-4′-cyanobiphenyl.

(b) Nematogens containing alicyclic rings

The first development in this area occurred through the work of Deutscher *et al.* (1977) who used the *trans*-1,4-disubstituted cyclohexane ring with its collinear but non-coaxial 1,4 bonds, substituting this for the benzene ring in the acid moiety of esters of the general structure shown at the bottom right of table 5. That is, ring A was changed from benzene (BZ) to cyclohexane (CH). This led to the development by Eidenschink *et al.* (1977) (Merck group, Darmstadt) of the cyclohexane analogues of the cyanobiphenyls: the *trans*-1-alkyl-4-(4′-cyanophenyl)cyclo-hexanes (the PCH series). Later, the 1,4-disubstituted bicyclo(2.2.2)octane ring was used by the Hull University group (Gray & Kelly 1980, 1981 *a, b*) in similar esters and cyanobicyclic systems to produce the compounds at the bottom of table 5, where ring A is bicyclooctane (BCO). The results in table 5 show that, for both types of system, T_{NI} falls from BCO to CH to BZ, i.e. the T_{NI} order is as shown at the top of figure 10.

This order has been verified now for over 20 structurally different systems (Gray 1981; Carr *et al.* 1981), so that it applies widely, and not only to cyano compounds and simple esters.

[15]

G. W. GRAY

However, for the cyano compounds shown at the foot of table 5, it has been demonstrated (D. A. Dunmur & A. E. Tomes, personal communication 1981; Ibrahim & Haase 1981) that the order of the anisotropy of the molecular polarizability ($\Delta\alpha$) is that shown in figure 10, and is not the same as the T_{NI} order.

T_{NI} order:
$$\text{BCO} > \text{CH} > \text{BZ} \quad \text{(20 systems)}$$

order of $\Delta\alpha$:
$$\text{BZ} > \text{BCO} > \text{CH} \quad \text{(cyanobiphenyl type)}$$

molecular polarisability

few exceptions   (to T_{NI} order):

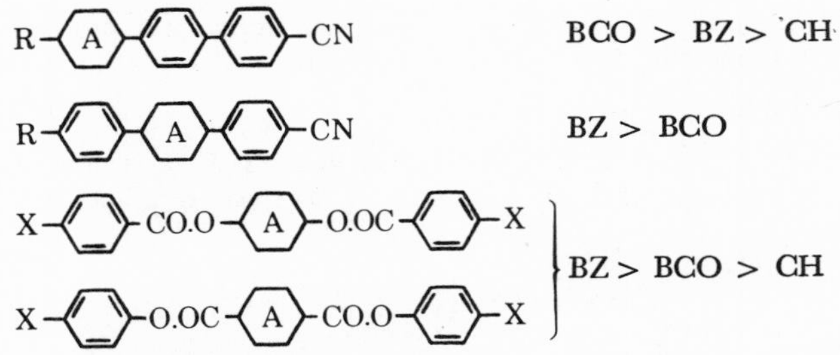

1 R—(A)—()—()—CN   BCO > BZ > CH

2 R—()—(A)—()—CN   BZ > BCO

3 X—()—CO.O—(A)—O.OC—()—X

4 X—()—O.OC—(A)—CO.O—()—X   BZ > BCO > CH

A = BZ

FIGURE 10. Data relating to the effects of cyclohexane and bicyclo(2.2.2)octane rings in the core structures of mesogens.

TABLE 5. USE OF ALICYCLIC RINGS

(1) *trans*-1,4-disubstituted cyclohexane

collinear
not coaxial

(2) 1,4-disubstituted bicyclo(2.2.2)octane

collinear
coaxial

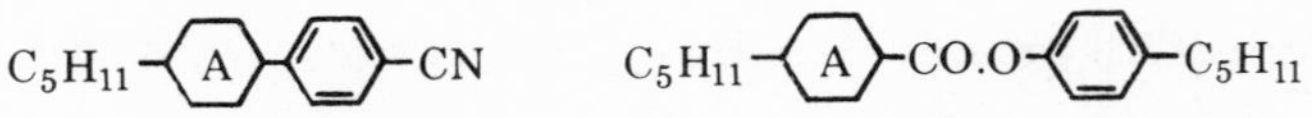

C_5H_{11}—(A)—()—CN C_5H_{11}—(A)—CO.O—()—C_5H_{11}

A	$T_{NI}/°C$	$T_{NI}/°C$
BCO	100	61.5
CH	55	48
BZ	35	26

We must therefore conclude that either the molecular interactions that determine T_{NI} are not solely a function of the anisotropic London dispersion forces, or the true significance of the London dispersion forces is not revealed by the experimental measurement of $\Delta\alpha$.

It is noted in figure 10 that some four exceptions to the general T_{NI} order are known. In the first, there is an inversion of the positions of BZ and CH; in the second, the position of CH is not clear, but BZ and BCO are inverted; in the last two, BZ rises above BCO > CH. The first may reflect differences in the extent of antiparallel overlap in the 'dimers' but, in the other three, $\Delta\alpha$ could be severely affected by the interruption of the extensive conjugated system that is possible when A = BZ.

Because the order BCO > CH > BZ cannot be explained in terms of $\Delta\alpha$ or antiparallel

[16]

correlation effects,† it is proposed that ring flexibility may be of significance. The cyclohexane ring is flexible, and this may assist in the achievement of an energetically economical close packing of the molecules – superior to that when ring A = BZ, but minimized to some extent by the non-coaxiality of the 1,4 bonds and the possibility that over-flexing of the ring can cause the 1,4 bonds to depart from strict collinearity through the adoption of skewed or twisted ring conformations. The BCO ring is also quite flexible, involving a twisting, concertina-like movement about the 1,4 bonds. Coupled with the bulk of the BCO ring, this could assist in efficient close packing of the molecules and efficient filling of space, giving a stable nematic state. Note that, for the BCO ring, the above motion of the ring does not lead to any deviation from collinearity of the 1,4 bonds. It is possible then that the BCO ring gives a particularly favourable combination of strict collinearity of its 1,4 bonds and sufficient flexibility of the ring structure, so favouring the nematic potential for a phase composed of molecules containing the ring, except in the few exceptional cases mentioned above.

TABLE 6

$$C_5H_{11} - \text{ring} - CO.O - \overset{X}{\text{ring}} - C_5H_{11}$$

X	$T_{CN}/°C$	$T_{NI}/°C$	$\Delta\epsilon$
H	31	64.5	−0.6
F	26	65	−0.95
Cl	35.5	41	—
Br	14.5	27	—
CN	27.5	29.5	−3.2

The BCO ring makes possible a wider range of structural possibilities.

It is interesting to record that the rules of thumb summarized earlier for aromatic nematogens can still be applied usefully for such systems. As shown in table 6, the bulk of the BCO ring shields the molecular broadening effect of 2-substituents in the phenolic moiety of the BCO esters of the structure shown (Gray & Kelly 1981 c). Whereas a 2-fluoro substituent substantially decreases T_{NI} in the corresponding esters derived from the benzoic acids and cyclohexanoic acids, in BCO esters the effect is minimal, and T_{NI} may increase by up to 2 K for some homologues. Thus $\Delta\epsilon$ can be made more negative without paying a significant penalty in T_{NI}. Larger groups such as 2-Cl, 2-Br and 2-CN *do* depress T_{NI}, but much less than in the benzoate and cyclohexanoate esters, so that such BCO esters still show enantiotropic nematic phases, with a strong negative $\Delta\epsilon$, particularly in the 2-cyano substituted BCO ester.

Such observations have led to the examination of other cycloaliphatic materials such as the cyclohexenyl compounds **1** and **2** in table 7. In compound **1** (Sato *et al.* 1979) the vinylic association of the double bond with the benzene ring promotes $\Delta\alpha$ (not incidentally considered to be unimportant, but just one factor in several), and this may contribute to the enhanced T_{NI} relative to the analogous PCH compound, **4**. In compound **2** (Osman & Revesz 1982) the stereochemistry of the molecule is adverse, and models show that an efficient close packing of molecules will not readily be achieved. Note that T_{NI} for compound **2** is monotropic, and this is conventionally indicated in parenthesis in table 7.

† PCH compounds and their BCO analogues are similarly antiparallel paired.

[17]

TABLE 7. CYCLOHEXENE SYSTEMS

1 C_7H_{15}—⬡—⬡—CN N $\xrightarrow{61\ ^\circ C}$ I

2 C_7H_{15}—⬡—⬡—CN (N $\xrightarrow{5\ ^\circ C}$ I)

3 C_7H_{15}—⬡—⬡—CN N $\xrightarrow{42\ ^\circ C}$ I

4 C_7H_{15}—⬡—⬡—CN N $\xrightarrow{59\ ^\circ C}$ I

(c) Nematogens containing cubane rings

Our own explorations of novel cycloaliphatic rings have included the cubane system, and, as will emerge, the results from this work reinforce the view that flexibility of rings and collinearity of bonds in the ring structures play a most significant role in determining T_{NI}. Although, as shown by the data in table 8, the dimensions of the cubane ring (strictly collinear 1,4 bonds) are in all ways intermediate between those of benzene and bicyclo(2.2.2)octane, the results shown in table 9 clearly illustrate (Gray *et al.* 1981) that replacement of a benzene or a cyclohexane or a bicyclooctane ring by a cubane ring in any system so far studied very markedly depresses T_{NI}. In fact, the cubane ring is a very bad unit to incorporate in the core structure of a mesogen.

TABLE 8. DIAMETERS OF CYLINDER OF ROTATION (ÅNGSTRÖMS)

	$C_1 \cdots C_4$	$H \cdots H$ (width)	$C \cdots C$ (width)
(benzene)	2.78	4.31	2.41
(cubane)	2.68	4.59	2.53
(bicyclooctane)	2.55	5.61	2.88

Cubane is always intermediate.

TABLE 9. C_3H_7-RING-CO.OY

(Examples from Gray *et al.* (1981).)

ring	Y	$T_{CN}/^\circ C$	$T_{NI}/^\circ C$
(benzene)	—⬡—C_5H_{11}	14	20
(bicyclooctane)	—⬡—C_5H_{11}	29	55
	—⬡—⬡—CN	142	289
	—⬡—⬡—C_3H_7	156	221
(cyclohexane)	—⬡—C_5H_{11}	32	(29)
	—⬡—⬡—CN	94	249
(cubane)	—⬡—C_5H_{11}	51	[−60]
	—⬡—⬡—CN	107	171
	—⬡—⬡—C_3H_7	101	116

[18]

Measurements of $\Delta\alpha$ for cubanes have not yet been made, but calculations of strain energy reveal that the high strain energy of the ring (650.8 kJ mol^{-1}) means that the cubane system is highly rigid. This strain energy is relieved if bonds in the cubane system are cleaved by hydrogenolysis, and it is very interesting to see that if the right bonds are successively cleaved, as shown in figure 11, we progress from cubane (**5**), to the dihydrocubane (**6**), to the tetrahydrocubane (**7**) and eventually obtain the hexahydrocubane (**8**), which is none other than bicyclo(2.2.2)octane. The data given in figure 11 also show that the strain energy is relieved in roughly equal amounts in the steps involved. As stated earlier, this emphasizes that the BCO

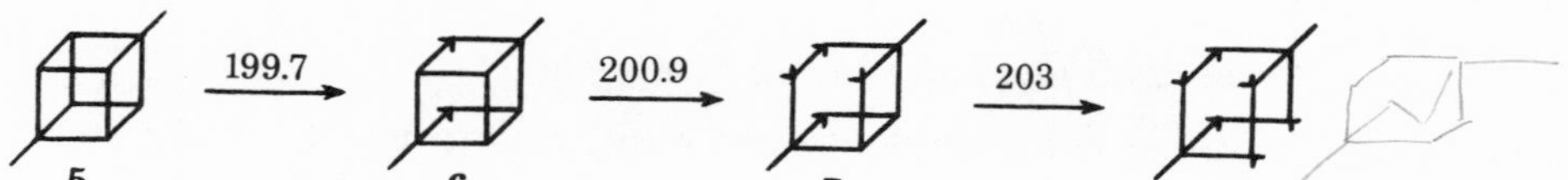

5 **6** **7** **8**

cubane hydrogenolysed cubanes

structure **8** is equivalent to

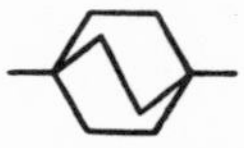

bicyclo(2.2.2)octane

above each arrow is the relief in strain energy (kilojoules per mole) associated with the particular hydrogenolysis step

1,4-bond angles: dihydrocubane, 22°; tetrahydrocubane, 12°; cubane and BCO, 0°.

FIGURE 11. Hydrogenolysis of the cubane system in three steps yields the bicyclo(2.2.2)octane system with progressive relief in strain energy.

ring is low in strain energy (47.2 kJ mol^{-1}) and quite flexible, and that the dihydrocubane and tetrahydrocubane rings might represent better core units than cubane itself, although the 1,4 bonds in the dihydrocubane and tetrahydrocubane do now deviate somewhat from collinearity (22° for the dihydrocubane and 12° for the tetrahydrocubane system). A balance will therefore be involved between the two factors, increased flexibility raising and non-collinearity depressing T_{NI}.

So far, only mesogens of the dihydrocubane system have been successfully prepared (K. J. Toyne *et al.*, unpublished results). Hydrogenolysis to the tetrahydrocubane ring has been successfully achieved, but rigid purification of an intermediate product required for the production of suitable mesogenic esters is still being carried out. However, as shown by the data in table 10, T_{NI} does increase on passing from a cubane ester to a dihydrocubane ester. Despite the nonlinearity of the bonds, therefore, the dihydrocubane ring is superior to cubane, and this could reflect the greater flexibility of the ring. Presumably this again gives a more efficient packing of the molecules with better space filling, resulting in stronger London dispersion forces than might be expected.

TABLE 10

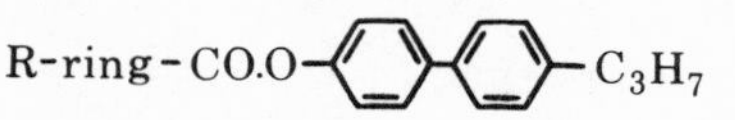

ring	T_{CN}/°C	T_{NI}/°C
cubane	110	118
dihydro	80	134
tetrahydro	?	?
BCO	131	232

As said earlier, it is felt that these results stress the need to take molecular flexibility and quite fine detail of stereochemistry into account, and it will be most interesting to see whether the tetrahydrocubane ring does in fact generate T_{NI} values intermediate between those of the dihydrocubane system and the known high values of the BCO analogues.

(d) Other cases of interest

In case the above discussion leads to the view that phase behaviour can readily be related to molecular structure, provided that thought be given to the situation, it is salutary to end with two examples that stress that we have still a great deal to learn.

(i) Comparisons of some cyano-substituted bicyclic nematogens

The cyano-substituted compounds under consideration here have the structures shown in figure 12 and the terminal substituent remote from the cyano group is in all cases an n-alkyl group with five carbon atoms, i.e. n-C_5H_{11}. The data include unpublished results by R. Eidenschink of E. Merck, Darmstadt, and D. Lacey and S. M. Kelly of Hull University.

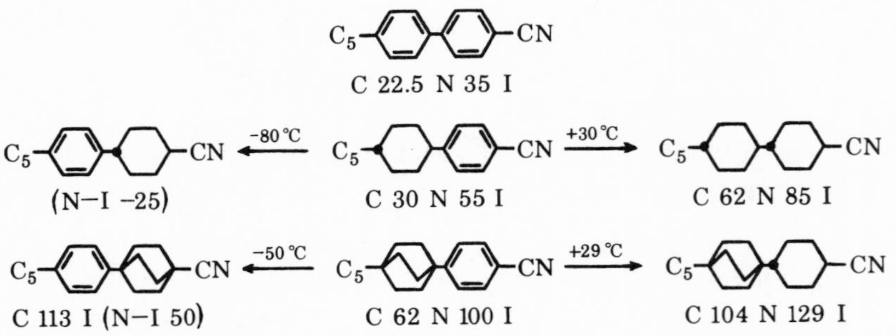

FIGURE 12. Structural variants of cyano-substituted bicyclic materials and the effects upon the nematic–isotropic liquid transition temperature (T_{NI}).

The three compounds in the central vertical block of figure 12 are the cyanobiphenyls, the PCH compounds and the corresponding BCO compounds, and the T_{NI} values show the normal increase on passing from BZ to CH to BCO. Moving to the right, we have the bicyclohexyl (CCH) and the bicyclooctylcyclohexyl analogues. The result of converting the ring carrying the cyano group from benzene to cyclohexane is a marked increase in T_{NI}. The effect could arise from the different antiparallel correlation situation in these alicyclic nitriles (only the cyano groups may be overlapped, giving a 'dimeric' situation). However, a similar situation could be argued for the compounds on the left side of figure 12 – also alicyclic nitriles – achieved by inverting the alicyclic ring and the aromatic ring in the appropriate compound in the central block. Now, however, the T_{NI} values are very low. Satisfactory explanations for these effects are difficult to find, but it is noted that:

(1) the isolated, unconjugated cyano group, with a strong localized dipole moment, is only well tolerated in the fully alicyclic systems;

(2) the highest T_{NI} values arise when the molecule can be divided into one high and one low polarizability part. Conversely, a sequence of low, high, low then high polarizability units – as in the left-hand structures – is apparently unfavourable to a stable nematic order.

[20]

(ii) *Mesogens with aliphatic ether functions*

Evidence accumulated (Gray 1981) from a range of systems has shown that whereas an alkoxy function linked to an aromatic nucleus favours a high T_{NI} value, an alkoxy group linked to alkyl or cycloalkyl gives a very low T_{NI} value. These conclusions have stemmed from studies of ethers whose molecules carry additional aromatic rings, and the first two examples (Osman 1982) recorded in table 11 show the large depressions in T_{NI} when the $-CH_2O-$ function is substituted for $-CO.O-$. However, the purely aliphatic ether – the third compound in table 11 – has a much more respectable T_{NI} value relative to the ester (Osman 1982). Therefore the isolated, unconjugated, dipolar ether function appears to be best tolerated in a purely alicyclic system, and this reminds us of conclusion (1) in §(d)(i) above. It therefore seems possible that, in both instances, interactions between aromatic rings and localized dipoles in neighbouring molecules may result either in repulsive interactions or in attractive forces that favour a non-ideal alignment for the formation of a nematic phase; in either case, the result is a lowering of the resistance of the nematic phase to the disaligning effects of increasing temperature.

TABLE 11

(Osman (1982).)

	$T_{NI}/°C$	T_{NI}(ester)/°C	$\Delta T/°C$
C$_3$H$_7$—⬡—CH$_2$O—⬡—⬡	(22)	(69.5)	−47.5
C$_5$H$_{11}$—⬡—⬡—CH$_2$O—⬡—C$_3$H$_7$	87.7	177.3	−89.6
C$_3$H$_7$—⬡—CH$_2$O—⬡—C$_3$H$_7$	17.5	36.6	−19.1

The purely alicyclic ether tolerates satC–O–satC quite well: compare satC–CN link in CCHs and BCOCHs.

CONCLUSIONS

Whatever the true explanations may be for some of the unusual effects described above, it is clear that studies of alicyclic mesogens have in recent years revealed a range of new factors that need to be taken into account when assessing the effects of molecular structural change upon liquid crystal transition temperatures, and theory must now move to accommodate the new facts.

The author wishes to express his gratitude to members (past and present) of his research group on whose work he has drawn for the purposes of this article. The names over the years would be many, but particular thanks for the more recent results quoted are due to Dr K. J. Toyne, Dr D. Lacey, Mr N. Carr and Mr S. M. Kelly.

REFERENCES

Brownsey, G. J. & Leadbetter, A. J. 1981 *J. Phys., Paris* **42**, L135–L139.
Bunning, J. D., Goodby, J. W., Gray, G. W. & Lydon, J. E. 1980 In *Liquid crystals of one- and two-dimensional order* (ed. W. Helfrich & H. Heppke), pp. 397–402. Berlin, Heidelberg and New York: Springer-Verlag.
Bunning, J. D., Lydon, J. E., Eaborn, C., Jackson, P. M., Goodby, J. W. & Gray, G. W. 1982 *J. chem. Soc. Faraday Trans. I* **78**, 713–724.
Carr, N., Gray, G. W. & Kelly, S. M. 1981 *Molec. Cryst. liq. Cryst.* **66**, 267–282.

de Jeu, W. H. 1977 *J. Phys., Paris* **38**, 1265–1273.

de Jeu, W. H., van der Veen, J. & Goossens, W. J. A. 1973 *Solid State Commun.* **12**, 405–407.

Demus, D. & Richter, L. 1978 In *Textures of liquid crystals*, ch. 4.10, pp. 91–92. Leipzig: V.E.B. Deutscher Verlag für Grundstoffindustrie.

Demus, D., Marzotko, D., Sharma, N. K. & Wiegeleben, A. 1980 *Kristall Technik* **15**, 331–339.

Deutscher, H.-J., Kuschel, F., König, S., Kresse, H., Pfeiffer, D., Wiegeleben, A., Wulf, J. & Demus, D. 1977 *Z. Chem.* **17**, 64–65.

Eidenschink, R., Erdmann, D., Krause, J. & Pohl, L. 1977 *Angew. Chem. int. Edn Engl.* **16**, 100.

Goodby, J. W., Gray, G. W. & McDonnell, D. G. 1977 *Molec. Cryst. liq. Cryst. Lett.* **34**, 183–188.

Gray, G. W. 1962 In *Molecular structure and the properties of liquid crystals*, pp. 1–314. London and New York: Academic Press.

Gray, G. W. 1976 In *Advances in liquid crystals* (ed. G. H. Brown), vol. 2, pp. 1–72. New York, San Francisco and London: Academic Press.

Gray, G. W. 1978 In *Advances in liquid crystal materials for applications*, pp. 1–42. Poole, Dorset: B.D.H. Chemicals Ltd.

Gray, G. W. 1981 *Molec. Cryst. liq. Cryst.* **63**, 3–18.

Gray, G. W. & Harrison, K. J. 1971*a* *Molec. Cryst. liq. Cryst.* **13**, 37–60.

Gray, G. W. & Harrison, K. J. 1971*b* *Symp. chem. Soc., Faraday Div.* no. 5, pp. 54–67.

Gray, G. W., Harrison, K. J. & Nash, J. A. 1973 *Electron. Lett.* **9**, 130–131.

Gray, G. W. & Kelly, S. M. 1980 *J. chem. Soc. chem. Commun.*, pp. 465–466.

Gray, G. W. & Kelly, S. M. 1981*a* *J. chem. Soc. Perkin Trans. II*, pp. 26–31.

Gray, G. W. & Kelly, S. M. 1981*b* *Angew. Chem.* **93**, 413–414.

Gray, G. W. & Kelly, S. M. 1981*c* *Molec. Cryst. liq. Cryst.* **75**, 109–119.

Gray, G. W., Langley, N. A. & Toyne, K. J. 1981 *Molec. Cryst. liq. Cryst. Lett.* **64**, 239–245.

Gray, G. W. & McDonnell, D. G. 1975 *Electron. Lett.* **11**, 556–557.

Ibrahim, I. H. & Haase, W. 1981 *Molec. Cryst. liq. Cryst.* **66**, 189–198.

Leadbetter, A. J., Richardson, R. M. & Colling, C. N. 1975 *J. Phys. Paris* **36**, 37–43.

Luzzati, V. & Spegt, P. A. 1967 *Nature, Lond.* **215**, 701–704.

Lydon, J. E. 1981 *Molec. Cryst. liq. Cryst. Lett.* **72**, 79–88.

McMillan, W. L. 1973 *Phys. Rev.* A **8**, 1921–1929.

Osman, M. A. 1982 *Molec. Cryst. liq. Cryst. Lett.* **82**, 47–52.

Osman, M. A. & Revesz, L. 1982 *Molec. Cryst. liq. Cryst. Lett.* **82**, 41–46.

Pohl, L., Eidenschink, R., Krause, J. & Weber, G. 1978 *Phys. Lett.* A **65**, 169–172.

Sato, H., Furukawa, K. & Sugimori, S. 1979 European Patent Disclosure, no. EP 0 002 136.

Tardieu, A. & Billard, J. 1976 *J. Phys., Paris* **37**, 79–81.

Tardieu, A. & Luzzati, V. 1970 *Biochim. biophys. Acta* **219**, 11–17.

Van der Meer, B. W. & Vertogen, G. 1979 *J. Phys., Paris* **40**, 222–228.

Phil. Trans. R. Soc. Lond. A **309**, 93–103 (1983)
Printed in Great Britain

Liquid crystals of disc-like molecules

By S. Chandrasekhar

Raman Research Institute, Bangalore 560080, India

Liquid crystals of disc-like molecules fall into two distinct structural types, the columnar and the nematic. The columnar phase in its simplest form consists of discs stacked aperiodically in columns, the different columns constituting a two-dimensional array, whereas the nematic phase is an orientationally ordered arrangement of discs without any long-range translational order. After a brief review of the structures of the various mesophases, two topics are considered in somewhat greater detail.

(i) An extension of McMillan's mean field model of smectic A to liquid crystals of disc-like molecules. The translational order is now assumed to be periodic in two dimensions. Calculations show that when the lattice is hexagonal, or departs from it only slightly, the transition from the columnar to the isotropic phase may take place either directly or via a nematic phase, depending on the model potential parameters α. Interpreting α to be a measure of the chain lengths as in McMillan's model, the theoretical phase diagram is shown to be in broad agreement with the experimental trends. As the asymmetry of the lattice is increased the theory predicts the occurrence of a smectic A phase as well. The new smectic A phase is biaxial.

(ii) A theoretical study of fluctuations in the columnar liquid crystal. It is shown that the Frank elasticity of the liquid-like columns stabilizes the two-dimensional order, a result that was in fact envisaged by Peierls and Landau in the 1930s. The mean-square fluctuation of the lattice as well as the Debye–Waller factor show a certain dependence on the linear dimensions of the sample.

Introduction

The shape of the molecule is an important criterion for determining liquid crystalline behaviour. Until quite recently, the accepted rule was that the molecule has to be elongated and rod-like for thermotropic mesomorphism to occur, but it has emerged in the last few years that *pure* compounds composed of relatively simple disc-like molecules may also form *stable* liquid crystals. The first example of this kind of mesomorphism was observed in the hexa-substituted esters of benzene. These compounds were prepared by my colleague B. K. Sadashiva at my instance, and from optical, thermodynamic and X-ray studies we concluded that they form an entirely new type of liquid crystal, quite unlike the classical nematic or smectic types that have been investigated for over 90 years. The structure that we proposed is illustrated in figure 1: the discs are stacked *aperiodically* in columns, the different columns constituting a two-dimensional array (Chandrasekhar *et al.* 1977). Thus the mesophase has translational periodicity in two dimensions but liquid-like disorder in the third. It has been variously designated as 'canonic' (from the Greek κανων = rod (Sir Charles Frank, personal communication 1978); see also Frank & Chandrasekhar (1980) and Helfrich (1980)), 'columnar' (Helfrich 1979) and 'discotic' (Billard *et al.* 1978), the last often being used to describe the molecules as well as the mesophases formed by them.

A number of other disc-like mesogens have since been found, notably by the Paris and Bordeaux groups (Dubois 1978; Billard *et al.* 1978; Destrade *et al.* 1979*a*), the canonic or columnar structure has been confirmed by the very fine X-ray work of Levelut (1979, 1980, 1982) and in particular, in the last 2 years or so, it has been established that these compounds exhibit a rich polymorphism, comparable with that observed in systems of rod-like molecules. The aim of the present paper is to review the current situation in the field. I shall first describe briefly what is known about the structures of these mesophases and then proceed to discuss some aspects of the physics of these new systems.

FIGURE 1. Schematic representation of the structure of the columnar liquid crystal.

BRIEF DESCRIPTION OF THE LIQUID CRYSTALLINE STRUCTURES

Figure 2 presents the chemical formulae of the disc-like mesogens reported to date. The mesophases so far discovered fall into two distinct categories, the columnar and the nematic.

The basic columnar structure (hereafter designated by the symbol D) is as shown in figure 1, but a number of variants have been identified: hexagonal, rectangular, tilted, etc. (figure $3a-f$). In all of them except one the columns are 'liquid-like', i.e. there is no long-range translational order along the axes of the columns. The one exception is the hexagonal phase of the hexa-n-alkoxytriphenylenes. From an X-ray study of the pentyl derivative, Levelut (1979) has inferred that there is a certain degree of regularity in the stacking of the triphenylene cores in each column, but that the chains are in a relatively disordered state and the molecular centres in neighbouring columns *uncorrelated*.

The nematic (N_D) phase is an orientationally ordered arrangement of the discs with no long-range translational order (figure $4a$). Unlike the usual nematic of rod-like molecules, the N_D phase is optically negative. A twisted nematic or cholesteric (N_D^*) phase has also been found (Destrade *et al.* 1980b; Malthete *et al.* 1981): compound (c) of figure 2 with the optically active chain R = S-($+$)CH$_3$-CH$_2$-CH(CH$_3$) (CH$_2$)$_3$O- exhibits such a phase, the helical configuration of which is depicted in figure 4b. For more detailed information on these different structures, the temperatures and entropies of transition, etc., reference may be made to recent review articles on the subject (Chandrasekhar 1982; Levelut 1982).

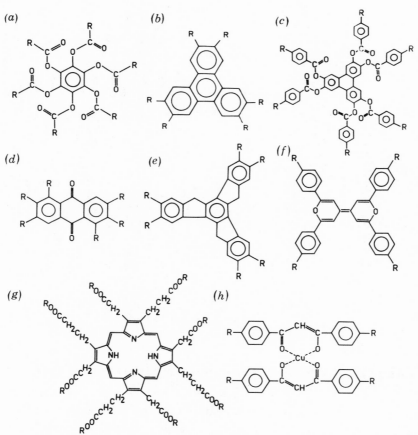

FIGURE 2. Disc-like mesogens: (*a*) hexa-*n*-alkanoates of benzene (Chandrasekhar *et al.* 1977, 1979); (*b*) hexa-*n*-alkanoates of triphenylene and hexa-*n*-alkoxytriphenylene (Dubois 1978; Billard *et al.* 1978; Destrade *et al.* 1979*a*, *b*); (*c*) hexa-*n*-alkyl and alkoxybenzoates of triphenylene (Tinh *et al.* 1979; Destrade *et al.* 1980*a*; Tinh *et al.* 1981); (*d*) rufigallol-hexa-*n*-octanoate (Queguiner *et al.* 1980; Billard *et al.* 1981); (*e*) hexa-*n*-alkanoates of truxene (Destrade *et al.* 1981*a*); (*f*) 2,2',6,6'-tetra-arylbipyran-4-ylidenes (Fugnitto *et al.* 1980); (*g*) uroporphyrin I octa-*n*-dodecyl ester (Goodby *et al.* 1980); (*h*) bis(*p-n*-decylbenzoyl)methanato copper(II) (Girod-Godquin & Billard 1981).

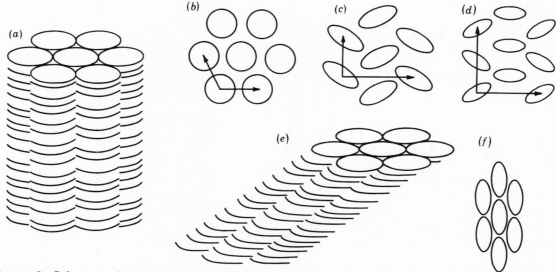

FIGURE 3. Columnar phases of disc-like molecules: (*a*) upright columnar structure; (*b*) its hexagonal and (*c*), (*d*) rectangular modifications; (*e*) tilted columnar structure; (*f*) its face-centred rectangular lattice.

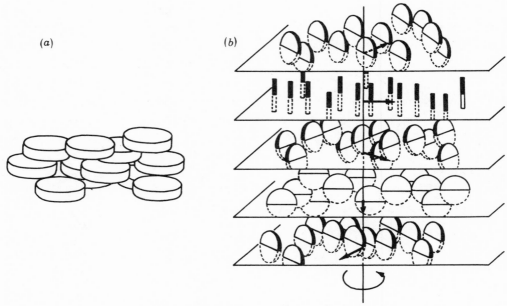

FIGURE 4. (a) The nematic phase N_D and (b) the twisted nematic or
cholesteric phase N_D^* of disc-like molecules.

THEORY OF THE COLUMNAR–NEMATIC–ISOTROPIC TRANSITIONS

Transitions between the D and N_D phases have been observed in a few compounds (Tinh
et al. 1981; Destrade *et al.* 1981 *a, b*). As an example, I present in table 1 the data for the hexa-*n*-
alkoxybenzoates of triphenylene, the structure of which is shown in figure 2*c* (Tinh *et al.* 1981).
Here D_t signifies tilted columns and a face-centred rectangular lattice (figure 3*e, f*), D_r upright
columns and a rectangular arrangement as in figure 3*c*. X-ray studies have shown that in the
latter structure the discs are *normal* to the columnar axes, implying thereby that there is an
asymmetric disposition of the chains about the molecular cores (Levelut 1980; Destrade *et al.*
1981 *b*). In both D_t and D_r the columns themselves are liquid-like. For the $n = 11$ homologue the
lattice parameters have been determined to be $a = 32.6$ Å and $b = 51.8$ Å (Tinh *et al.* 1981;
Destrade *et al.* 1981*b*), which represents only a very slight departure from true hexagonal
symmetry.

TABLE 1. HEXA-*n*-ALKOXYBENZOATES OF TRIPHENYLENE:
TRANSITION TEMPERATURES IN DEGREES CELSIUS

$R = C_n H_{2n+1} O$	K		D_t		D_r		N_D		I
$n = 4$	.	257	—		—		.	> 300	.
5	.	224	—		—		.	298	.
6	.	186	.	193	—		.	274	.
7	.	168	—		—		.	253	.
8	.	152	—		.	168	.	244	.
9	.	154	—		.	183	.	227	.
10	.	142	—		.	191	.	212	.
11	.	145	—		.	179	.	185	.
12	.	146	—		.	174	—		.

K, crystal; D_t, tilted columnar; D_r, rectangular columnar; N_D, nematic; I, isotropic. The phases exhibited by
a compound are indicated by points in the appropriate columns.

It is seen from table 1 that the lower members of the homologous series, $n = 4$ and 5, show only the N_D phase. The next few members (with the exception of $n = 7$) show both D and N_D phases, the temperature range of the N_D phase decreasing with increasing chain length till at $n = 12$ the D phase transforms directly to the isotropic (I) phase. Broadly the trend is reminiscent of the behaviour of the smectic A–nematic–isotropic (S_A–N–I) transitions in systems of rod-like molecules. This suggests that one may be able to give a qualitative description of the D–N_D–I transitions by extending McMillan's mean field model of S_A (McMillan 1971) so that the density wave is now periodic in two dimensions (Kats 1978; Feldkamp *et al.* 1981; Chandrasekhar *et al.* 1982).

Consider a face-centred rectangular lattice composed of liquid-like columns, the molecular cores being assumed to be circular discs normal to the columnar axes. Such a lattice can be described by a superposition of three density waves with wavevectors (figure 5)

$$A = 2\pi(i/a + j/b);$$
$$B = 2\pi(-i/a + j/b);$$

and
$$C = 4\pi j/b = (A + B).$$

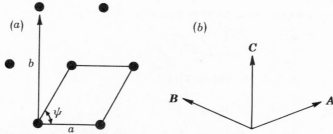

FIGURE 5. Two-dimensional face-centred rectangular lattice showing (*a*) the lattice parameters *a* and *b* and the primitive cell of the direct lattice, (*b*) the reciprocal lattice vectors *A*, *B* and *C*.

Because orientational ordering is obviously necessary for the existence of the columnar structure, we couple each density wave to the appropriate component of the orientational order parameter along *A*, *B* or *C*. Retaining only the leading terms in the Fourier expansions of the density waves, the single particle potential in the mean field approximation may be taken to be of the form (Chandrasekhar *et al.* 1982)

$$V_1(r, \theta, \phi) = -V_0[-\eta_1 P_2(\sin\theta\cos\phi) - \eta_2 P_2(\sin\theta\sin\phi)$$
$$- 2\alpha_1\sigma_1 P_2(\sin\theta\cos\phi)\cos(C\cdot r) - 2\alpha_2\sigma_2\{P_2(\sin\theta\cos(\psi+\phi))\cos(A\cdot r)$$
$$+ P_2(\sin\theta\cos(\psi-\phi))\cos(B\cdot r)\}],$$

where V_0 determines the nematic–isotropic transition temperature T_{NI}, α_2 is an interaction strength related to the density waves along *A* and *B* and α_1 to the wave along *C*, $\psi = \tan^{-1}(b/a)$, θ and ϕ are the polar angles, *r* is the position vector, and P_2 is the Legendre polynomial of order two. This form of the potential ensures that the energy of the molecule is minimum when the disc is centred in the column with its plane normal to the *z* axis.

Because the molecular core is assumed to be circularly symmetric, we get

$$\alpha_1 = 2\exp\{-(2\pi r_0/b)^2\},$$
$$\alpha_2 = 2\exp[-\{\pi r_0(a^2+b^2)^{\frac{1}{2}}/ab\}^2],$$

where r_0 is the range of the intermolecular attractive potential.

[27]

In general there are four order parameters, η_2 and η_1, the orientational order parameters measured along the x and y axes, and σ_2 and σ_1, the order parameters coupling the orientational and the translational ordering along the A (or B) and C directions respectively. We have chosen to define the order parameters in such a way that they vary from 0 in the disordered system to $+1$ in the perfectly ordered system. We then have

$$\eta_1 = \langle -2P_2(\sin\theta\cos\phi)\rangle,$$

$$\eta_2 = \langle -2P_2(\sin\theta\sin\phi)\rangle,$$

$$\sigma_1 = \langle -2P_2(\sin\theta\cos\phi)\cos(C\cdot r)\rangle,$$

$$\sigma_2 = \langle -[P_2\{\sin\theta\cos(\psi+\phi)\}\cos(A\cdot r)+P_2\{\sin\theta\cos(\psi-\phi)\}\cos(B\cdot r)]\rangle,$$

where the angular brackets represent a statistical average, and the normalized single partition distribution function

$$f_1(r,\theta,\phi) = \frac{\exp\{-V_1(r,\theta,\phi)/k_B T\}}{\int dr \int_0^1 d(\cos\theta) \int_0^{2\pi} d\phi \exp\{-V_1(r,\theta,\phi)/k_B T\}},$$

where $\int dr$ is over the primitive cell, and k_B is the Boltzmann constant. The molar internal energy of the oriented system can now be written as

$$\frac{\Delta U}{Nk_B T} = -\frac{V_0}{2k_B T}(\tfrac{1}{2}\eta_1^2 + \tfrac{1}{2}\eta_2^2 + \alpha_1\sigma_1^2 + 2\alpha_2\sigma_2^2),$$

where N is the Avogadro number, the entropy as

$$\frac{\Delta S}{Nk_B} = -\int_0^1\int_0^{2\pi}\int f(r,\theta,\phi)\ln f(r,\theta,\phi)\,d(\cos\theta)\,d\phi\,dr$$

$$= -\frac{V_0}{k_B T}(\tfrac{1}{2}\eta_1^2 + \tfrac{1}{2}\eta_2^2 + \alpha_1\sigma_1^2 + 2\alpha_2\sigma_2^2) + \ln\frac{1}{\pi ab}\int_0^1\int_0^{2\pi}\int\exp\{-V_1(r,\theta,\phi)/k_B T\}d(\cos\theta)\,d\phi\,dr,$$

and hence the molar free energy as

$$\frac{\Delta F}{Nk_B T} = \frac{\Delta U - T\Delta S}{Nk_B T} = \frac{V_0}{2k_B T}(\tfrac{1}{2}\eta_1^2 + \tfrac{1}{2}\eta_2^2 + \alpha_1\sigma_1^2 + 2\alpha_2\sigma_2^2)$$

$$-\ln\frac{1}{\pi ab}\int_0^1\int_0^{2\pi}\int\exp\{-V_1(r,\theta,\phi)/k_B T\}d(\cos\theta)\,d\phi\,dr.$$

There are four possible solutions to the equations:

(1) $\eta_1 \neq \eta_2 \neq 0$, $\sigma_1 \neq \sigma_2 \neq 0$ (biaxial rectangular columnar phase);
(2) $\eta_1 \neq \eta_2 \neq 0$, $\sigma_1 \neq 0, \sigma_2 = 0$ (biaxial smectic phase);
(3) $\eta_1 = \eta_2 \neq 0$, $\sigma_1 = \sigma_2 = 0$ (uniaxial nematic phase);
(4) $\eta_1 = \eta_2 = \sigma_1 = \sigma_2 = 0$ (isotropic phase).

When $b/a = \sqrt{3}$, $\psi = 60°$, and we have a hexagonal lattice. It then follows that $\alpha_1 = \alpha_2$ and the solutions take the simpler form

(1) $\eta_1 = \eta_2 \neq 0$, $\sigma_1 = \sigma_2 \neq 0$ (uniaxial hexagonal columnar phase);
(2) $\eta_1 = \eta_2 \neq 0$, $\sigma_1 = \sigma_2 = 0$ (uniaxial nematic phase);
(3) $\eta_1 = \eta_2 = \sigma_1 = \sigma_2 = 0$ (isotropic phase).

The free energy corresponding to the different solutions can be evaluated to determine the phase diagram as a function of the α coefficients for a given value of the axial ratio b/a.

The phase diagram for the hexagonal structure is shown in figure 6. The hexagonal–nematic transition is always first-order, as is to be expected. It is seen that the temperature range of the nematic phase decreases with increasing α, and for $\alpha > 0.64$ the columnar phase transforms directly to the isotropic phase. If, as in McMillan's theory, α is interpreted to be a measure of the chain length these results are in qualitative accord with the observed trends (table 1).

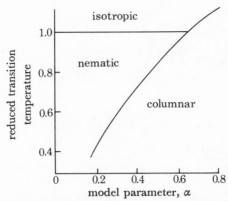

FIGURE 6. Theoretical plot of the reduced transition temperatures against the model parameter α showing the hexagonal, nematic and isotropic phase boundaries. All the transitions are of first order.

For the biaxial rectangular structure, the nature of the phase diagram depends on b/a. I shall confine my remarks to $b/a > \sqrt{3}$. If b/a is only slightly greater than $\sqrt{3}$, the phase diagram is similar to that for the hexagonal structure. As the asymmetry of the lattice is increased, one gets solutions corresponding to a smectic phase: the rectangular columnar phase transforms to a biaxial smectic A phase (with the layer normal along C) and this in turn undergoes a transition to a uniaxial N_D phase at a higher temperature (Chandrasekhar *et al.* 1982). Rectangular phases have been observed with molecules having essentially circular cores, but in none of them does b/a depart very much from $\sqrt{3}$. It remains to be seen whether rectangular lattices of sufficiently high anisotropy, high enough to give rise to a smectic phase, will be found. Nevertheless, our simple model serves to illustrate the fact that the origin of the two-dimensional translational order in the columnar phase is similar to that of the one-dimensional order in the smectic A phase of rod-like molecules in so far as the attraction between the aromatic cores and the role of the end chains are concerned.

FLUCTUATIONS IN THE COLUMNAR LIQUID CRYSTAL

I shall next consider the fundamental question of fluctuations in the columnar phase. This and the related problem of stability were discussed in general terms by Peierls and Landau (see Landau & Lifshitz 1970), and have recently been investigated by Ranganath & Chandrasekhar (1982) and independently and rather more thoroughly by Kammensky & Kats (1982).

Let us suppose that the liquid-like columns are along the z axis and that the two-dimensional (2D) lattice (assumed to be hexagonal) is parallel to the xy plane. We shall consider only the vibrations of the lattice in the xy plane. The free energy density may be expressed as

$$F = \tfrac{1}{2}B\left(\frac{\partial u_x}{\partial x}+\frac{\partial u_y}{\partial y}\right)^2 + \tfrac{1}{2}D\left\{\left(\frac{\partial u_x}{\partial x}-\frac{\partial u_y}{\partial y}\right)^2 + \left(\frac{\partial u_x}{\partial y}+\frac{\partial u_y}{\partial x}\right)^2\right\} + \tfrac{1}{2}k_{33}\left\{\left(\frac{\partial^2 u_x}{\partial z^2}\right)^2 + \left(\frac{\partial^2 u_y}{\partial z^2}\right)^2\right\}, \qquad (1)$$

[29]

where B and D are the elastic constants for the deformation of the 2D lattice in its own plane, u_x and u_y are the displacements along x and y at any lattice point, and k_{33} is the Frank constant for the curvature deformation (bending) of the columns. (In the notation of the standard crystal elasticity theory, $B = \frac{1}{2}(c_{11}+c_{12})$ and $D = \frac{1}{2}(c_{11}-c_{12})$). We neglect here the splay and twist deformations because they give rise to a distortion of the lattice (Prost & Clark 1980). We also neglect any contributions from the surface of the sample. Writing the displacement u in terms of its Fourier components,

$$u(r) = \sum_q u(q) \exp[iq \cdot r]; \qquad (2)$$

substituting in (1) we get in the harmonic approximation

$$F = \frac{1}{2} \sum_q (B_0 q_\perp^2 + k_0 q_z^4) \langle u_q^2 \rangle,$$

and from the equipartition theorem

$$\langle u_q^2 \rangle = k_B T/(B_0 q_\perp^2 + k_0 q_z^4),$$

where $B_0 = B+2D$, $k_0 = 2k_{33}$ and $q_\perp = (q_x^2+q_y^2)^{\frac{1}{2}}$.

The mean square displacement at any lattice point is given by

$$\langle u^2 \rangle = \sum_q \langle u_q^2 \rangle = \frac{1}{(2\pi)^3} \int \langle u_q^2 \rangle \, dq$$

$$= \frac{k_B T}{(2\pi)^3} \int_{2\pi/L}^{2\pi/d} \int_{2\pi/L'}^{\infty} \frac{2\pi q_\perp \, dq_\perp \, dq_z}{B_0 q_\perp^2 + k_0 q_z^4},$$

where L' is the length of the columns, L the linear dimension of the lattice in the xy plane and d its periodicity. Assuming that $L' \gg L$

$$\langle u^2 \rangle = \{k_B T/4B_0(\lambda d)^{\frac{1}{2}}\} \{1 - (d/L)^{\frac{1}{2}}\}, \qquad (3)$$

where $\lambda = (k_0/B_0)^{\frac{1}{2}}$ is a characteristic length (Ranganath & Chandrasekhar 1982; Kammensky & Kats 1982). The structure is therefore stable as $L \to \infty$ (see Landau & Lifshitz 1970). As is well known (Peierls 1934; Landau 1967) the 2D lattice itself is an unstable system with $\langle u^2 \rangle$ diverging as $\ln L$ (Jancovici 1967; Gunther et al. 1980). Thus the curvature elasticity of the liquid-like columns stabilizes the 2D order in the columnar liquid crystal.

The 'structure factor' for the intensity of X-ray scattering may be written as

$$S(K) = \int dz \sum_m \sum_n \exp[i\{K \cdot (R_m - R_n)\}] \exp[-\frac{1}{2}K_\perp \langle |u_n - u_m|^2 \rangle].$$

The second exponential term on the right hand side is the familiar Debye–Waller factor $\exp(-W)$. Now

$$\langle (u_n - u_m)^2 \rangle = \langle |u(r) - u(0)|^2 \rangle$$

$$= \frac{2}{2\pi^3} \int \langle u_q^2 \rangle |1 - \cos(q \cdot r)| \, dq \quad \text{(from (2))}$$

$$= \frac{k_B T}{2B_0(\lambda d)^{\frac{1}{2}}} \left\{ 1 - \left(\frac{d}{L}\right)^{\frac{1}{2}} \right\} - \frac{k_B T}{16\sqrt{2\pi}B_0(\lambda\rho)^{\frac{1}{2}}} \exp\left(-\frac{z^2}{4\lambda\rho}\right) \left\{ U\left(\frac{1}{4}, \frac{1}{2}, \frac{z^2}{4\lambda\rho}\right) \right\}^2, \qquad (4)$$

where $\rho = (x^2+y^2)^{\frac{1}{2}}$ and U is the confluent hypergeometric Kummer function (see Abramovitz & Stegun 1965). For $z \gg (\lambda\rho)^{\frac{1}{2}}$, the second term of (4) reduces to

$$\frac{\sqrt{2} k_B T}{16\pi B_0 z} \exp\left(-\frac{z^2}{4\lambda\rho}\right).$$

[30]

In smectic A and the 2D lattice, the displacement–displacement correlation is of logarithmic form; the Bragg peaks (δ–function singularities) are therefore washed out for the infinite sample and there appears instead a strong thermal diffuse scattering with weaker singularities (Caillé 1972; Gunther *et al.* 1980; Litster 1980; Als-Nielsen 1981). On the other hand, the columnar liquid crystal does give the usual Bragg reflexions. Preliminary calculations show that the Debye–Waller factor is determined mainly by the first term of (4). Now, the lattice spacing d in any of the known columnar phases does not exceed about 5 nm, so that the size effect in (3) and (4) is quite weak in practical situations. With larger disc-like mesogens, the effect may well be observable.

I have so far considered the case of long columns ($L' \gg L$). If the columns are short ($L' \ll L$) and the sample is bounded (normal to the columnar axis on both sides), $\langle u^2 \rangle$ is independent of L' for very small L', and varies as $|\text{const.} - (\ln L'/L')|$ for larger L'. If the surfaces are free $\langle u^2 \rangle$ will have additional terms, which may be analogous to those for a 2D lattice. In any case, as far as X-ray scattering is concerned, such a thin free film may be expected to behave differently from a sample in the form of a narrow cylinder consisting of long columns.

REFERENCES

Abramovitz, M. & Stegun, I. A. 1965 In *Handbook of mathematical functions*, p. 504. New York: Dover.

Als-Nielsen, J. 1981 In *Symmetries and broken symmetries* (ed. N. Boccara), pp. 107–122. Paris: Idset.

Billard, J., Dubois, J. C., Tinh, N. H. & Zann, A. 1978 *Nouv. J. Chim.* **2**, 535–540.

Billard, J., Dubois, J. C., Vaucher, C. & Levelut, A. M. 1981 *Molec. Cryst. liq. Cryst.* **66**, 115–122.

Caillé, A. 1972 *C.r. hebd. Séanc. Acad. Sci., Paris*, B **274**, 891–893.

Chandrasekhar, S. 1982 In *Advances in liquid crystals* (ed. G. H. Brown), vol. 5, pp. 47–78. New York and London: Academic Press.

Chandrasekhar, S., Sadashiva, B. K. & Suresh, K. A. 1977 *Pramana* **9**, 471–480.

Chandrasekhar, S., Sadashiva, B. K., Suresh, K. A., Madhusudana, N. V., Kumar, S., Shashidhar, R. & Venkatesh, G. 1979 *J. Phys., Paris* **40**, c3-120–c3-124.

Chandrasekhar, S., Savithramma, K. L. & Madhusudana, N. V. 1982 In *Proceedings of the American Chemical Society Symposium on Ordered Fluids and Liquid Crystals, Las Vegas, March* 1982 (ed. J. F. Johnson & A. C. Griffin). New York and London: Plenum Press. (In the press.)

Destrade, C., Bernard, M. C., Gasparoux, H., Levelut, A. M. & Tinh, N. H. 1980*a* In *Proceedings of the International Conference on Liquid Crystals, Bangalore, December* 1979 (ed. S. Chandrasekhar), pp. 29–32. London, Philadelphia and Rheine: Heyden.

Destrade, C., Gasparoux, H., Babeau, A., Tinh, N. H. & Malthete, J. 1981*a* *Molec. Cryst. liq. Cryst.* **67**, 37–48.

Destrade, C., Mondon, M. C. & Malthete, J. 1979*a* *J. Phys., Paris*, **40**, c3-17–c3-21.

Destrade, C., Mondon, M. C. & Tinh, N. H. 1979*b* *Molec. Cryst. liq. Cryst. Lett.* **49**, 169–174.

Destrade, C., Tinh, N. H., Malthete, J. & Jacques, J. 1980*b* *Physics Lett.* A **79**, 189–192.

Destrade, C., Tinh, N. H., Gasparoux, H., Malthete, J. & Levelut, A. M. 1981*b* *Molec. Cryst. liq. Cryst.* **71**, 111–135.

Dubois, J. C. 1978 *Annls Phys.* **3**, 131–138.

Feldkamp, G. E., Handschy, M. A. & Clark, N. A. 1981 *Physics Lett.* A **85**, 359–362.

Frank, F. C. & Chandrasekhar, S. 1980 *J. Phys., Paris* **41**, 1285–1288.

Fugnitto, R., Strzelecka, H., Zann, A., Dubois, J. C. & Billard, J. 1980 *J. chem. Soc. chem. Commun.*, pp. 271–272.

Girod-Godquin, A. M. & Billard, J. 1981 *Molec. Cryst. liq. Cryst.* **66**, 147–150.

Goodby, J. W., Robinson, P. S., Teo, B. K. & Cladis, P. E. 1980 *Molec. Cryst. liq. Cryst. Lett.* **56**, 303–309.

Gunther, L., Imry, Y. & Lajzerowicz, J. 1980 *Phys. Rev.* A **22**, 1733–1740.

Helfrich, W. 1979 *J. Phys., Paris* **40**, c3-105–c3-114.

Helfrich, W. 1980 In *Proceedings of the International Conference on Liquid Crystals, Bangalore, December* 1979 (ed. S. Chandrasekhar), pp. 7–19. London, Philadelphia and Rheine: Heyden.

Jancovici, B. 1967 *Phys. Rev. Lett.* **19**, 20–22.

Kammensky, V. G. & Kats, E. I. 1982 Preprint of the Landau Institute of Theoretical Physics, Moscow.

Kats, E. I. 1978 *Soviet Phys. JETP* **48**, 916–920.

Landau, L. D. 1967 In *Collected papers of L. D. Landau* (ed. D. Ter Haar), pp. 210-211. New York, London and Paris: Gordon & Breach.

Landau, L. D. & Lifshitz, E. M. 1970 *Statistical physics*, 2nd edn, pp. 402–404. Oxford: Pergamon Press.

Levelut, A. M. 1979 *J. Phys. Lett., Paris* **40**, L81–L84.
Levelut, A. M. 1980 In *Proceedings of the International Conference on Liquid Crystals, Bangalore, December* 1979 (ed.
    S. Chandrasekhar), pp. 21–27. London, Philadelphia and Rheine: Heyden.
Levelut, A. M. 1982 *J. Chim. phys.* (In the press.)
Litster, J. D. 1980 In *Proceedings of the Conference on Liquid Crystals, of One- and Two-dimensional Order, Garmisch-
    Partenkirchen, F.R.G., January* 1980 (ed. W. Helfrich & G. Heppke), pp. 65–70. Berlin, Heidelberg and
    New York: Springer-Verlag.
Malthete, J., Destrade, C., Tinh, N. H. & Jacques, J. 1981 *Molec. Cryst. liq. Cryst. Lett.* **64**, 233–238.
McMillan, W. L. 1971 *Phys. Rev.* A **4**, 1238–1246.
Peierls, R. E. 1934 *Helv. phys. Acta Suppl.* **7**, 81–83.
Prost, J. & Clark, N. A. 1980 In *Proceedings of the International Conference on Liquid Crystals, Bangalore, December*
    1979 (ed. S. Chandrasekhar), pp. 53–58. London, Philadelphia and Rheine: Heyden.
Queguiner, A., Zann, A., Dubois, J. C. & Billard, J. 1980 In *Proceedings of the International Conference on Liquid
    Crystals, Bangalore, December* 1979 (ed. S. Chandrasekhar), pp. 35–40. London, Philadelphia and Rheine:
    Heyden.
Ranganath, G. S. & Chandrasekhar, S. 1982 *Curr. Sci.* **51**, 605–606.
Tinh, N. H., Destrade, C. & Gasparoux, H. 1979 *Physics Lett.* A **72**, 251–254.
Tinh, N. H., Gasparoux, H. & Destrade, C. 1981 *Molec. Cryst. liq. Cryst.* **68**, 101–111.

Discussion

Sir Charles Frank, F.R.S. (*The University, Bristol*). Professor Chandrasekhar said that with an infinite smectic sample the Bragg reflexions would disappear. Now, I grant that any such negative statement can be made about an experiment with an infinite sample, since such an experiment cannot be performed, but I presume the intended statement is about the limiting behaviour approached as the specimen is made larger and larger; and says that that reflexion does not progressively sharpen as it would with a true crystal, rather than that it disappears.

J. D. Litster. The reflexion will not disappear; it will still be rather strong. However, the line shape will be a power-law singularity instead of the delta function Bragg peak one normally observes from a system with long-range order.

Sir Charles Frank. I think that makes it a matter of semantics: wouldn't the experimentalist still say he had a Bragg reflexion?

J. D. Litster. Unless the resolution function of his spectrometer was sufficiently good to distinguish the two, he would. However, the difference is important; it is whether one has long-range order or not.

W. H. de Jeu (*Solid State Physics Laboratory, Groningen, The Netherlands*). Recently I learned that the hexa-substituted benzenes, the first compounds to show discotic mesophases, were synthesized some 50 years ago at the University of Groningen (H. J. Backer & S. van der Baan, *Rec. Trav. Chim. Pays-Bas* **56**, 1161–1174 (1937)). More interestingly they appeared to be still in stock in the Department of Organic Chemistry. We studied the compound with $O–CO–C_6H_{13}$ as substituent and observed indeed the expected columnar mesophase between 80 and 86 °C. Powder X-ray photographs (Guinier-Simon camera with $CoK\alpha_1$ radiation) showed the reflexions listed in table D 1.

The first two reflexions are strong, the others rather weak. Indexing is possible on the basis of a 2D orthorhombic lattice of the columns with $a = 28.5$ Å and $b = 17.7$ Å. The lattice is thus not hexagonal (in which case $a/b = \sqrt{3}$), although the deviations from hexagonal are not large ($a/b = 1.61$).

TABLE D 1

hkl	$d_{obs}/\text{Å}\dagger$	$d_{calc}/\text{Å}\dagger$
110	15.0	15.04
200	14.3	14.27
210	11.1	11.11
020	8.86	8.85
120	8.45	8.45

$\dagger$ $1\text{ Å} = 10^{-10}\text{ m} = 10^{-1}\text{ nm}$.

S. CHANDRASEKHAR. In our very first report on these compounds (Chandrasekhar et al. 1977) we had noted that some X-ray photographs contained a few weak diffraction spots not quite conforming to true hexagonal symmetry, but the quality of the X-ray patterns did not warrant a more refined analysis. Subsequently, from optical observations of the extinction brushes in relatively large domains in which the columnar axes are curved in circles or spirals, Sir Charles Frank showed that the lattice should be pseudohexagonal (Frank & Chandrasekhar 1980). This has recently been confirmed by Madame A. M. Levelut (to be published in *J. Chim. phys.*), who found that the lattice is rectangular with an axial ratio departing by about 3% from the true hexagonal value of $\sqrt{3}$.

LIN LEI (*Institute of Physics, Chinese Academy of Sciences, Beijing, China*). We have modified the McMillan hamiltonian so that in the two-particle interaction $W_{ij} = W_0(r_{ij}) + W_2(r_{ij}) P_2(\cos\theta_{ij})$, both W_0 and W_2 are functions of the molecular length d. For $W_0 = F(d) V_0(r_{ij})$, $W_2 = F(d) V_2(r_{ij})$ and $F(d) = d^\lambda$, our results for the I–N–A phase diagrams of long molecules reduce essentially to those of McMillan (1971) and Lee et al. (1973) when $\lambda = 1$ and 0 respectively (corresponding to T_{IN} being constant or decreasing with n, the number of carbon atoms in the end-chain of the molecule). For $\lambda = 2$, we find T_{IN}^- increasing with n; for more general $F(d)$, it is possible to have a minimum in T_{IN} (Shu Changqing & Lin Lei, *Acta phys. sin.* **31**, 915 (1982)). This simple method of obtaining different types of $T_{IN}(n)$ curve should also be applicable in Professor Chandrasekhar's theory in which disc-like molecules are treated.

Phil. Trans. R. Soc. Lond. A **309**, 105–114 (1983)
Printed in Great Britain

Liquid crystalline side-chain polymers

By H. Finkelmann

Physikalisch Chemisches Institut der Technischen Universität Clausthal,
3392 Clausthal-Zellerfeld, Adolf-Romer-Strasse 2A, F.R.G.

Liquid crystalline side-chain polymers are polymers that have linked conventional low molecular mass liquid crystals (l.c.) as side chains to a polymer backbone. Their properties are experimentally compared with the corresponding low molecular mass l.c. (i) Beginning with a monomeric l.c., with increasing degree of polymerization an increase of the phase transformation temperature of l.c. to isotropic is observed. (ii) The nematic order parameter of the monomers is about 10 % higher than the order parameter of the polymers. Owing to the linkage of the rigid mesogenic molecules to the polymer backbone, their rotation around the long molecular axis differs strongly from that of the monomers. (iii) Owing to the high viscosity of the polymers their response times in the electric field are much larger than the response times of the monomers. If a sufficient length of the flexible spacer that connects the mesogenic molecule to the polymer main chain is assumed, the threshold voltages of the monomers and polymers are of comparable magnitude.

1. Introduction

During the last few years, polymers exhibiting a liquid crystalline state have become of increasing theoretical and technological interest because of their properties, which are a combination of polymer-specific properties and the anisotropic behaviour of liquid crystals (l.cs). Like the monomeric l.c. (m.l.c.) the polymeric l.c. (p.l.c.) can be differentiated by their chemical constitution (figure 1).

If we look at the monomer units of the polymer backbone, in most cases we can identify these units as having the same chemical constitution as m.l.cs. Three different structures are known: (i) non-amphiphilic cylindrical, (ii) non-amphiphilic disc-like and (iii) amphiphilic monomer units. To form a polymeric main chain these mesogenic elements can be connected in two different ways: (i) head to tail, forming 'l.c. main-chain polymers', and (ii) head to head, resulting in 'l.c. side-chain polymers' (figure 1).

While polymers consisting of disc-like monomers have not yet been achieved, the four remaining polymer types are known. By analogy with m.l.cs, thermotropic phase behaviour is almost only observed for the non-amphiphilic polymers.

In this paper I shall focus on the l.c. side-chain polymers. In their chemical constitution these polymers consist of two elements. One element is the polymer backbone, which can be widely varied in chemical and physical properties. The other element is the mesogenic side chain, for which also a large number of chemical constitutions are conceivable. One principal aspect of these polymers is whether polymer-specific properties are influenced by the anisotropic state of order of the mesogenic side chains and vice versa.

Compared with the monomeric l.c., in principle the only change of the mesogenic moiety is the restriction of translational and rotational motions due to the linkage to the backbone. These restrictions should be more or less influenced by the physical properties of the main chain and in which way the rigid mesogenic moiety is linked to the backbone.

[35]

With some experimental results in this paper I shall discuss the influence of the degree of polymerization on the l.c. phase behaviour of the polymers. Thereafter the state of order of m.l.cs will be compared with that of the chemically very similar polymers. Finally the behaviour of the polymers in an electric field will be compared with the behaviour of m.l.cs.

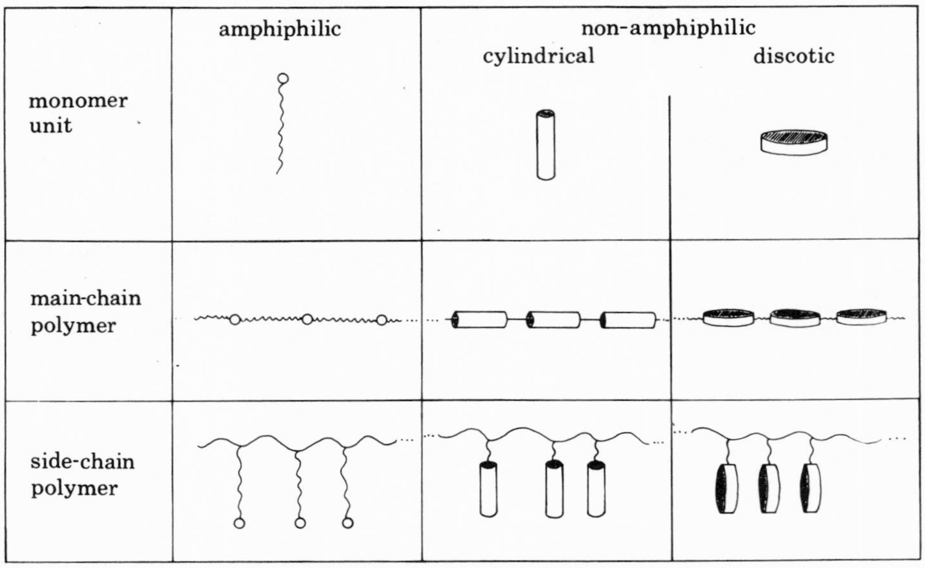

FIGURE 1. Scheme of liquid crystalline polymers.

2. DEGREE OF POLYMERIZATION

During the past few years it has been proved that nematic, cholesteric and smectic l.c. side-chain polymers can be made, exhibiting a thermodynamically homogeneous l.c. state (Finkelmann 1982). Starting from an l.c. monomer it has always been observed that because of the polymerization the extent of the l.c. phase of the monomer is strongly different from that of the polymer. To get a detailed insight into these mechanisms, Stevens $et\ al.$ (1983) have investigated the phase behaviour as function of the degree of polymerizations of polysiloxanes:

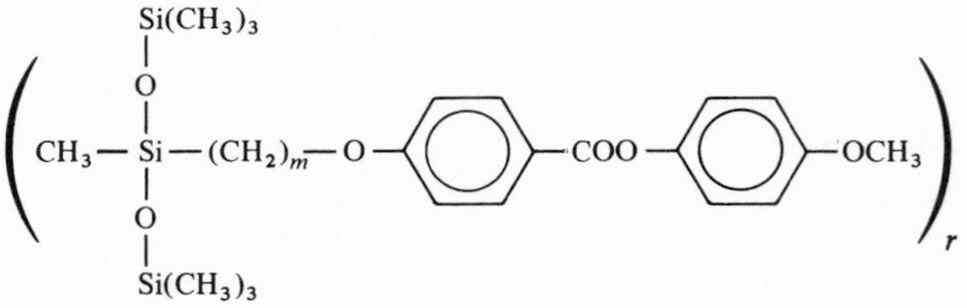

A mixture of oligomers with different degrees of polymerization, r, were separated by gel permeation chromatography into monodisperse oligomers with defined r. The phase behaviour of these defined oligomers as function of r is shown in figure 2 for oligomers with spacer lengths $m = 3$ and $m = 6$. Polymers with $m = 3$ exhibit a nematic phase and polymers with $m = 6$ have additionally a low-temperature smectic phase. For both systems the same characteristic behaviour is observed. While the derivatives with $r = 1$ and $r = 2$ are not (or only metastable) liquid crystalline, for $r > 3$ the phase transformation temperatures from nematic to isotropic, T_{NI}, increase sharply with r and remain nearly constant for $r \approx 10$. The same behaviour is also

[36]

observed for the transformation temperature T_{SN}, but r does not affect the transformation temperatures as much as for T_{NI}.

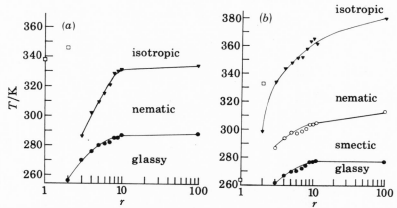

FIGURE 2. Phase behaviour of monodisperse oligomers as a function of the degree of polymerization, r: (a), $m = 3$; (b), $m = 6$. Squares: phase transformation crystalline to isotropic.

The change of the l.c. phase transformation temperatures towards higher temperatures with increasing r has been observed for all monomer–polymer systems investigated so far and can be established as a principal rule. Very often a change from nematic to the higher-ordered smectic phases is also observed. This 'stabilization' of the l.c. state by polymerization can be easily understood by the restriction of translational and rotational motions of the mesogenic molecules when they are linked to the polymer backbone.

Another important aspect resulting from the formation of the polymer main chain can be performed from figure 2. While the monomer and dimer ($r = 1$ or 2) crystallize, for $r > 3$ a crystallization is no longer observed at low temperatures but there is instead a transition into the glassy state. With increasing r the glass transition temperature also increases. This is well known for polymers. For this transition it should be noted that the structure of the l.c. phase observed at higher temperatures is converted into the glassy state without any macroscopically recognizable changes. This will be discussed in more detail in the next section.

3. Degree of order

On the basis of polarizing microscopic observations, nematic and cholesteric l.c. side-chain polymers qualitatively behave very similarly to m.l.cs. Textures and optical character (nematic phases, uniaxial positive; cholesteric phases, uniaxial negative) do not differ. On the other hand it is of interest whether the state of order is influenced when an m.l.c. is linked to a polymeric backbone. If a cylindrical symmetry of the mesogenic moieties is assumed, the state of order of the nematic phase can be described by the orientational order of the mesogenic molecules with respect to their long molecular axis $S = \frac{3}{2}(\cos^2\theta - \frac{1}{3})$, where θ is the angle between the long molecular axis and the director. The director is the symmetry axis of the orientational distribution function of the long molecular axis. In case of the polymers we refer S not to the whole macromolecule but to the mesogenic side chains of the monomer units of the macromolecule. By this way we neglect the polymer backbone. In figure 3 the chemical constitution is shown for three different systems, which will be compared with respect to S. S has been determined by Benthack *et al.* (1983) for a dye probe which is (*a*) dissolved in a monomer (M1), (*b*) dissolved in a polymer (P1) and (*c*) linked to the polymer (P2). The results are shown in figure 4. It is obvious that the m.l.c. and the p.l.cs show the same temperature dependence of S. At T_{NI} ($T^* = 1$), S vanishes discontinuously, indicating the first

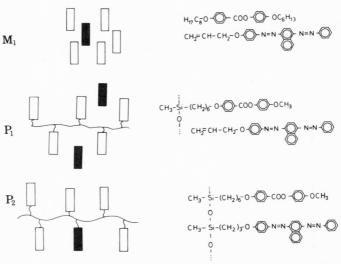

FIGURE 3. Chemical constitution of nematic phases incorporating a dye probe: M_1, a mixture of m.l.c. and dye; P_1, a mixture of p.l.c. and dye; P_2, copolymer of p.l.c. and dye.

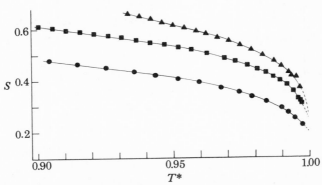

FIGURE 4. Temperature dependence of the order parameter, S, for three systems described in figure 3 ($T^* = T_m/T_{c1}$; T_m is measuring temperature; T_{c1} is phase transformation temperature, nematic to isotropic): ▲, M_1; ■, P_1; ●, P_2.

order phase transformation. Looking at a defined reduced temperature, T^*, we find that the dye dissolved in the monomer (M1) exhibits the highest-order parameter. For the dye dissolved in the polymer (P1) the order parameter is reduced for about 10%, which indicates that the polymer backbone disturbs the order of the nematic host. Birefringence measurements of S confirm these results. Furthermore it could be shown that S is not affected if the same mesogenic group is linked by flexible spacers of different lengths to the polymer main chain. An interesting effect is observed when the voluminous dye probe is also linked to the backbone in the copolymer (P2). Here the order of the dye is smaller than in the solution (P1). The decrease in S can be understood because of the direct linkage of the dye to the backbone. Owing to this linkage the dye cannot adapt the order of the host phase in the same way as in solution.

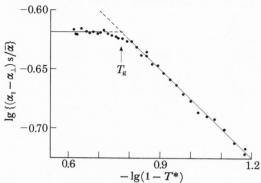

FIGURE 5. Temperature dependence of the order parameter, S, for a nematic polymer (Haller plot: $\alpha_\parallel$, $\alpha_\perp$ are molecular polarizabilities parallel and perpendicular to the molecular axis; $\bar{\alpha}$ = mean polarizability, for T^* refer to figure 4).

In figure 4 the order parameter S is shown for the polymers in the l.c. state. As mentioned before, in contrast to most m.l.cs the polymers very often exhibit at low temperatures a glass transition, which does not affect the optical properties of the system macroscopically. More detailed information is obtained if the temperature dependence of S is analysed passing through the glass transition. Following Haller *et al.* (1973), the graph of lg S against lg $(1 - T^*)$, where $T^* = T_m/T_c$ (T_m being the measuring temperature and T_c the phase transformation temperature l.c.–isotropic), yields a straight line. This is demonstrated for a polymer in figure 5 from birefringence measurements.

In figure 5 the ordinate is the logarithm of $S\,\Delta\alpha/\bar{\alpha}$, which is directly proportional to S. As expected for high temperatures, the straight line indicates a normal change of S with temperature. At low temperatures, however, a bend in the curve indicates a change in the function $S(T)$. The straight line becomes parallel to the abscissa, which means that S is constant and no longer changes with temperature. By thermodynamic measurements the bend in the curve can be attributed to the glass transition. These optical measurements prove that not only the texture but also the nematic order freezes in at T_g. With these properties interesting applications are conceivable, e.g. the production of optical storage material.

For the nematic order parameter, the orientational distribution of the long molecular axis with respect to the director is considered by assuming a cylindrical symmetry of the mesogenic molecules or the mesogenic moieties linked as side chains to the backbone. At least for chiral molecules, however, no cylindrical molecular shape can be assumed. Following the theory of Goossens (1979), additional order parameters have to be considered, which also include the rotation of the mesogenic moieties around their long molecular axis.

An experimental method, which directly reflects on the rotation of the mesogenic molecules around their long molecular axis, is the induced twist observed if chiral molecules are dissolved in a nematic host phase. The induced twist depends directly on the concentration of the chiral guest molecules, their chemical constitution and on an order parameter S_r, which describes their rotation around the long molecular axis. If there is a free rotation ($S_r = 0$), no cholesteric twist will be induced. With increasing hindered rotation an increasing twist will be obtained.

A change of the rotation around the long molecular axis of an m.l.c. molecule should occur on varying the temperature. A much stronger effect on S_r, however, should be obtained, if the rigid mesogenic molecule is linked by different flexible alkyl chains to a polymer backbone. Whereas for the monomeric l.c. molecule their motions are only restricted by the anisotropic

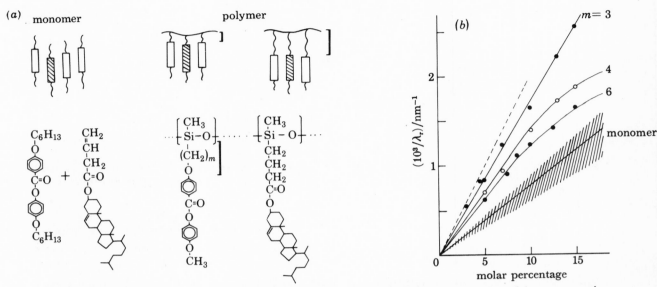

FIGURE 6. Inverse wavelength of reflexion, $1/\lambda_r$, of the induced cholesteric phase as a function of the concentration of chiral guest molecules (m is the length of the flexible spacer).

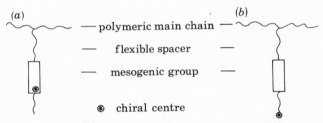

FIGURE 7. Scheme of chiral molecules with a chiral centre (a) within the rigid mesogenic group, and (b) linked by a flexible linkage to the rigid mesogenic group.

interactions with the neighbours, the motions of the mesogenic molecules, which are linked to a polymer backbone, are additionally restricted by this covalent linkage. With increasing length of the flexible spacer from the backbone to the rigid mesogenic moiety, increasing mobility and therefore a falling S_r and a falling induced twist are expected. This has been experimentally established (Finkelmann & Rehage 1980).

In figure 6b the inverse wavelength of reflexion, $1/\lambda_r$, of induced cholesteric phases (figure 6a), which are induced by chiral molecules, is shown as function of the concentration (in molar percentages) of the chiral guests. The inverse wavelength $1/\lambda_r$ is directly proportional to the induced twist.

For a constant concentration of the chiral guest the highest twist is observed if the mesogenic moieties are linked by a flexible spacer with three methylene groups ($m = 3$). With increasing length of the spacer, the twist becomes smaller. The lowest value of $(1/\lambda_r)_x$ is observed for the monomeric mixture.

To prove that the considerations of hindered rotation of the rigid mesogenic moieties are correct and that the experiments actually reflect this restriction of motions, we performed another experiment. This is presented schematically in figure 7. For the experiments mentioned above the chiral centres are within the rigid mesogenic part of the molecules (figure 7a). If, however, the asymmetric centre is also linked by a flexible linkage to the rigid mesogenic groups (figure 7b) a change in a hindered rotation of the mesogenic moiety should no longer

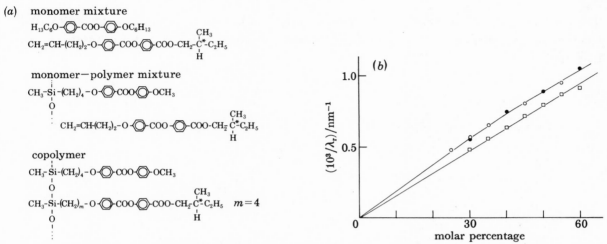

FIGURE 8. Inverse wavelength of reflexion, $1/\lambda_r$, of induced cholesteric phases as a function of the concentration of chiral guest molecules for: □, monomer mixture, ○, monomer–polymer mixture, and ●, copolymers.

strongly influence the chiral nematic twist. In this way it has to be presumed that the motions of the chiral groups are no longer correlated with respect to different kinds of linkages of the mesogenic moiety to the polymer main chain. These considerations are also confirmed by experiment (figure 8). If the monomer mixtures are compared with the polymer systems, only a small change in the induced twist can be observed, which might also be attributed to a change in the basic nematic order parameter. However, *no* difference is observed in the twist (i) for a mixture of the nematic polymer and the chiral monomer, (ii) for the copolymers with the nematic host and chiral guest molecules, and (iii) for the copolymers with different spacer lengths.

These experimental results are of great interest in view of the theories of the cholesteric phase and the basic question of whether or not a hindered rotation of the molecules has to be considered. For our experiments, in principle the chemical constitution of the systems (e.g. the summation formula), which are compared, remains almost constant. Only the motions of the mesogenic moieties are changed by different kinds of flexible linkages to the polymer backbone. Following our experiments, the results are only consistent with the ideas of the theory of Goossens (1979), in which the cholesteric twist is explained by the hindered rotation of the chiral molecules around their long molecular axis.

4. BEHAVIOUR IN AN ELECTRIC FIELD

During the past few years m.l.cs have become of great commercial interest because of their applicability to display technology. A mushrooming number of new substances has been prepared, to optimize the physical properties of the material for their use in the different kinds of display cells. Here three basic parameters are of interest: the dielectric anisotropy, $\Delta\epsilon$, the elastic constants, k_{ii} and the viscosity, approximately represented by the bulk viscosity η. These parameters determine

(i) the threshold voltage of the cells,

$$U_0 \propto (k_{ii}/\Delta\epsilon)^{\frac{1}{2}}, \qquad (1)$$

and (ii) the response time of the display,

$$\tau_{\text{on/off}} \propto \eta. \qquad (2)$$

The l.c. side-chain polymers are a completely new type of l.c. material and it is obvious to ask whether the polymers can be used in display technology or whether they bring prospects of new applications.

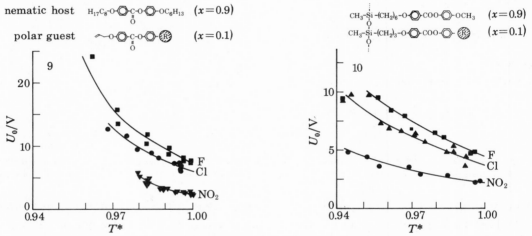

FIGURE 9. Temperature dependence of the threshold voltage of mixtures of monomers ($T^* = T_m/T_{NI}$).
FIGURE 10. Temperature dependence of the threshold voltage of copolymers.

Without doing any experiments we know that one property of the polymeric material strongly differs from that of the conventional l.c.: the viscosity of the polymer is larger by several orders of magnitude than that of the monomers. Therefore, referring to (2), we have to conclude that the response times of *pure* polymers are larger by several orders of magnitude than those of m.l.cs. Experiments confirm these considerations. If, however, the glass transition temperature T_g of the polymers is much smaller than the temperature of the experimental conditions, response times $\tau < 200$ ms have been observed (Finkelmann *et al.* 1979). With falling temperature, τ becomes infinite at T_g. To get a more detailed insight into the field effects of the polymers, Kiechle *et al.* (1983) compared the threshold voltage, U_0, of monomeric l.c. mixtures with chemically very similar copolymers in a TN display (sample thickness 15 μm, frequency 10 kHz).

In figure 9 the temperature dependence of U_0 is shown for monomer mixtures differing in the polarity of the substituent R of the polar guest molecule. According to (1), with increasing $\Delta\epsilon$ we observe a falling U_0. The same behaviour is observed for polymers with a spacer length of $m = 6$ (figure 10). If we compare the threshold voltage at a defined T^* we even observe $U_{0,\text{m.l.c.}} > U_{0,\text{p.l.c.}}$. As the chemical constitution of both systems, and thus also U_0, is very similar, the elastic constants of the m.l.c. and the p.l.c. must be of the same magnitude. Referring to the measurements of the order parameters (§3), polymers with a spacer length $m = 6$ behave more similarly with regard to m.l.cs than polymers with $m < 6$. The rigid mesogenic, groups are less hindered by the polymer backbone than are short flexible spacers. This behaviour is also reflected by the threshold voltage, which is compared for polymers with different spacer

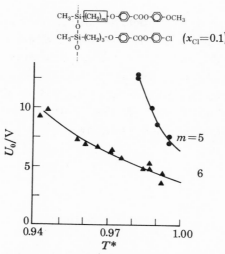

FIGURE 11. Threshold voltage U_0 as a function of the reduced temperature, T^*, for copolymers with different spacer lengths, m.

lengths but the same chemical constitution (figure 11). Comparing U_0 at a constant T^* the threshold voltage increases strongly with decreasing spacer length and for $m = 4$ we found $U_0 > 50$ V. The copolymer with $m = 3$ did not show any field effects under the experimental conditions. Following the arguments mentioned above, we can conclude that the elastic constants of the polymers depend on the chemical constitution of the linkage of the mesogenic groups to the polymer backbone.

To summarize these results, an application of pure l.c. side-chain polymers in fast switching displays is not practicable. On the other hand, a characteristic property of the polymers is the glassy state and the possibility to freeze an l.c. texture in the glassy state. With this property it is possible to produce films. Above the glass transition temperature, an electric field can be used to put in information that can be durably stored at lower temperatures in the glassy state of the material.

REFERENCES

Benthack, H., Finkelmann, H. & Rehage, G. 1983 *J. Phys., Paris.* (In the press.)
Finkelmann, H. 1982 In *Polymer liquid crystals* (ed. A. Ciferri, W. R. Krigbaum & R. B. Meyer), ch. 2, p. 35. New York: Academic Press.
Finkelmann, H. & Rehage, G. 1980 *Makromol. Chem. rapid Commun.* **2**, 317.
Finkelmann, H., Ringsdorf, H. & Naegele, D. 1979 *Makromol. Chem.* **180**, 803.
Goossens, W. J. A. 1979 *J. Phys., Paris Colloq.* (Orsay, France), vol. 40, p. 158.
Haller, J., Huggins, H. A., Lilienthal, H. R. & McGuire, T. R. 1973 *J. phys. Chem.* **77**, 950.
Kiechle, U., Finkelmann, H. & Rehage, G. 1983 *Molec. Cryst. liq. Cryst.* (In the press.)
Stevens, H., Finkelmann, H., Lühmann, B. & Rehage, G. 1983 In *Liquid crystals and ordered fluids* (ed. J. S. Johnson & R. S. Porter), vol. 4. (In the press.)

Discussion

A. H. Price (*Edward Davies Chemical Laboratories, University College of Wales, Aberystwyth, U.K.*). Would Dr Finkelmann please indicate how the response times to an external stimulus vary between an 'ordinary' nematogen, a side-chain liquid crystal polymer and a main-chain polymer?

H. Finkelmann. As described in the paper, for an l.c. side-chain polymer the response time depends strongly on the vicinity of a glass transition, where a field effect can no longer be observed. With increasing distance to T_g, the response time falls and for some polymers $\tau < 200$ ms has been found at high temperatures (200 °C).

For l.c. main-chain polymers the response time depends on the degree of polymerization, r. For high molecular mass polymers, no field effects were observed. Detailed information about the response time as a function of r is not yet known.

Phil. Trans. R. Soc. Lond. A **309**, 115–126 (1983)

Printed in Great Britain

Theories of nematic order

By T. E. Faber

Cavendish Laboratory, Madingley Road, Cambridge CB3 0HE, U.K.

The paper starts with a description of the molecular alignment that characterizes nematic liquid crystals, and a brisk review of the classic mean field theories of this alignment and of recent attempts to improve on the mean field approach by introducing the direct correlation function. It is argued that such theories are unsatisfactory on two grounds: (a) because they fail to recognize that correlations of orientation between adjacent molecules are of very much longer range than the intermolecular potential – they fall off, indeed, only as fast as $1/R$ – and (b) because they fail to allow for director fluctuations. The author's continuum theory, which attributes entirely to director fluctuations the fact that molecules in nematics are not perfectly aligned, is free from these particular objections, and it seems to give the most complete description currently available, especially at low temperatures, for the behaviour of simple model nematics that have been studied by computer simulation. Its failure to match completely the behaviour of real nematics such as 5CB may be due to their polar character.

1. Description of nematic order

In nematic liquid crystals the molecules are preferentially aligned with respect to a local axis of symmetry, known as the director. Since the symmetry is in practice uniaxial (though nematics that are locally biaxial are not difficult to imagine) the degree of alignment may be described for rigid molecules by two order parameters

$$S_{zz} = \langle P_2(\cos\beta)\rangle = \langle \tfrac{1}{2}(3\cos^2\beta - 1)\rangle \tag{1}$$

and

$$(S_{xx} - S_{yy}) = \langle \tfrac{3}{2}\sin^2\beta \cos 2\gamma\rangle. \tag{2}$$

In such expressions the angled brackets indicate an ensemble average for molecules in the neighbourhood of interest, and α, β and γ are Euler angles describing the orientation with respect to the director of (x, y, z) axes fixed in each molecule. The molecular axes might be chosen, for example, to coincide with the principal axes for the moment of inertia tensor for a single molecule, I_{ij}. If so, the conventional labelling (not wholly rational!) would be such as to ensure that

$$I_{yy} > I_{xx} > I_{zz}.$$

Thus the z axis is the 'long' axis of the molecule, and β is the angle between this and the director, while γ is the angle through which the molecule is rotated about its long axis, starting from an orientation such that the director lies in the xz plane. It seems that, given this labelling, $(S_{xx} - S_{yy})$ may in practice be of order $+0.05$ (Emsley *et al.* 1981) and large enough to affect some bulk properties, such as magnetic susceptibility (Faber *et al.* 1983). The parameter S_{zz} is of order $+0.5$, however, and is evidently of much greater significance. In what follows we shall lose sight of $(S_{xx} - S_{yy})$, supposing the molecules to rotate with sufficient freedom about their long axes to be treated as cylinders. We shall also lose sight of the complications that arise with molecules that are not rigid.

The parameter S_{zz} is sometimes denoted by S_2 instead, to emphasize that it is one member of the hierarchy of order parameters that are needed to define the full orientational distribution function, $f(\beta)$. We have

$$f(\beta) = \sum_{n=0}^{n} (2n+1) S_n P_n(\cos\beta),\tag{3}$$

with

$$S_n = \langle P_n(\cos\beta)\rangle.\tag{4}$$

It is invariably assumed that S_n vanishes when n is odd, because nematics whose molecules carry an electric dipole moment show no signs of ferroelectric ordering. As regards even n, it is fair to say that for most nematics we know only S_0 ($=1$) and S_2; the experimental evidence relating to S_4 is limited and not wholly reliable, except in the case of a few model nematics whose properties have been investigated by the 'experimental' technique of computer simulation.

Nematic order is not completely described by $f(\beta)$, even when the rigid cylinder approximation is sound, for this function tells us nothing about correlations in position and alignment for neighbouring molecules. To describe the correlations in position that characterize a simple isotropic liquid we need, of course, a two-body distribution function $g(R)$. The equivalent distribution function in a nematic needs to be written as

$$g(\boldsymbol{R}_{12}, \alpha_1, \beta_1, \alpha_2, \beta_2) = g(1,2)f(\beta_1)f(\beta_2)\tag{5}$$

(an equation that serves to define the function $g(1,2)$ to which reference is made below). In principle $g(1,2)$ describes correlations of orientation as well as of position, but if we wish to focus attention on the former it may be simpler to discuss what have been termed the short-range order parameters

$$\sigma_n(\boldsymbol{R}_{12}) = \langle P_n(\cos\theta_{12})\rangle, \quad n = 2, 4, \ldots,\tag{6}$$

where θ_{12} is the angle between the z axes of two molecules whose centres are separated by $\boldsymbol{R}_{12}$. Of course, the correlations must disappear, i.e. $g(1,2)$ must tend to unity, for large separations, and it is readily shown that in that limit

$$\sigma_2 = S_2^2.\tag{7}$$

The difference between σ_2 and S_2^2 provides a convenient measure of correlations of orientation at smaller separations.

Finally, we should note that in a macroscopic sample of nematic the director does not necessarily point in the same direction throughout. Various types of distortion, distinguished by the terms splay, twist and bend, may be fed into the director field by distorting the boundary conditions of the sample, at some cost in free energy. The cost is determined by the three Frank stiffness constants, k_{11}, k_{22} and k_{33}. Even when the boundary conditions are consistent with uniform alignment, some distortion always arises in the interior of the sample as a result of fluctuations, and it is for this reason that nematic liquids scatter light as strongly as they do.

2. Mean field and other theories

Attempts to understand the occurrence of nematic order have mostly (though by no means exclusively) been based upon the pioneering work of either Onsager (1949) or Maier & Saupe (1958, 1959, 1960). Since developments of Onsager's theory are likely to receive attention elsewhere in this symposium, we may concentrate initially on the Maier–Saupe approach. Its essence may be understood by considering a model rather simpler than the one that Maier &

Saupe initially proposed, i.e. a model in which the intermolecular pair potential contains an anisotropic term of short range, of the form

$$V(1, 2) = -\epsilon(\boldsymbol{R}_{12})\, P_2(\cos\theta_{12}). \tag{8}$$

Let us evaluate the free energy for this model, making the approximation that correlations of orientation can be completely ignored, i.e. that $g(1, 2)$ depends only on $\boldsymbol{R}_{12}$. This approximation allows one to express the entropy of misalignment as $-Nk\langle \ln f(\beta)\rangle$ and it also allows one to make use of (7) in evaluating the internal energy due to the interaction that (8) describes. The order-dependent terms in the free energy therefore turn out to be

$$-CS_2^2 + NkT\langle \ln f(\beta)\rangle, \tag{9}$$

where C should depend only on density. A simple variational argument (de Gennes 1974) shows that to minimize the free energy we need

$$f(\beta) \propto \exp\{-2CS_2 P_2(\cos\beta)/kT\}, \tag{10}$$

and a self-consistency relation then serves to fix S_2, once C and T are known. Like the Weiss theory of ferromagnetism, which it closely resembles, the theory of Maier & Saupe is a mean field one. The mean field in question is $-2CS_2$ and it couples with $P_2(\cos\beta)$.

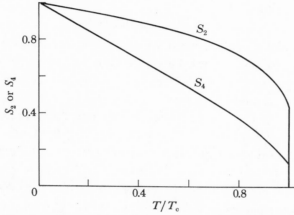

FIGURE 1. The dependence of S_2 and S_4 on T/T_c according to Maier & Saupe.

At constant density the Maier–Saupe theory involves only one adjustable parameter, the constant C in (9) or, if preferred, the temperature T_c at which the nematic phase becomes isotropic on heating. It therefore predicts unique curves, reproduced in figure 1, for the variation of S_2 and S_4 with T/T_c. The agreement with experiment is in some respects quite impressive, but it is certainly not complete. Several attempts have therefore been made to improve the theory, by adding terms to the intermolecular pair potential while retaining the mean field framework (Chandrasekhar & Madhusudana 1971, 1973; Humphries et al. 1972; Luckhurst et al. 1975; Shen et al. 1981). At the expense of one or more additional adjustable parameters, the fit to experiment can indeed be improved.

Unfortunately, the basic assumption that $g(1, 2)$ is a function only of $\boldsymbol{R}_{12}$ is not borne out by the results of computer simulation studies. Admittedly these studies have so far been restricted to rather simple models; the most detailed results available are for a $10 \times 10 \times 10$ lattice of 'molecules', with periodic boundary conditions, which interact via an anisotropic potential of

[47]

the sort that (8) describes, ϵ being zero except between nearest neighbours (Zannoni 1979). But if the correlations of orientation between neighbouring molecules are strong in such a lattice, as they seem to be, they are presumably strong in real nematics too.

It has been suggested (e.g. by Luckhurst & Zannoni (1977)) that the mean field arguments of Maier & Saupe should be used to discuss the alignment of *clusters* of molecules rather than of individual molecules, and that the analysis of the mean molecular structure within a cluster should be treated as a separate problem, to be solved exactly as far as possible. The approach is superficially attractive, though to go much beyond the theory of Sheng & Wojtowicz (1976), who considered clusters containing only two molecules, would evidently require much labour.

An alternative approach is that of Stecki & Kloczkowski (1979). In some ways this is more closely related to the theory of Onsager than to that of Maier & Saupe, but the results can be presented in a mean-field-like form, to the extent that an entropic term $NkT\langle\ln f(\beta)\rangle$ can be isolated from the free energy. The residue is expressed, exactly, in terms of a direct correlation function $c(1,2)$, which bears the same relation to $g(1,2)$ that $c(R)$ bears to $g(R)$ for a simple fluid, and similar functions of higher order, $c_3(1,2,3)$ and so on. In simple fluids various approximations are available to link $c(R)$ to the intermolecular pair potential $V(R)$, namely

$$C \approx \exp(-\beta V) - 1, \tag{11}$$

$$C \approx -\beta V, \tag{12}$$

and
$$C \approx g(R)\{1 - \exp(\beta V)\}, \tag{13}$$

with $\beta = 1/kT$. The second of these is used in the so-called mean spherical model of simple liquids and the third corresponds to the Percus–Yevick model. Stecki and his coworkers use similar equations to relate $c(1,2)$ to $V(1,2)$ in cases where the potential is anisotropic and where the molecules are perhaps aligned. They have been careful to restrict their calculations so far to circumstances where $c_3(1,2,3)$, etc., can be neglected. This restriction means that while they supply expressions for the change in free energy due to small fluctuations in S_2 about its equilibrium value, or due to gradual changes in the orientation of the director, they do not attempt to calculate curves for S_2 and S_4 in the nematic phase to replace those plotted in figure 1. Other authors have been less cautious (Ruijgrok & Sokalski 1982; Wagner 1981).

It is clear that Stecki's approach is capable in principle of describing correlations of orientation between nearest neighbours. Thus if $V(1,2)$ is such as to favour parallel alignment of molecules 1 and 2, it follows from (11), (12) or (13) that $c(1,2)$, and hence $g(1,2)$, will be enhanced for pairs that are parallel (θ_{12} small) at the expense of pairs that are non-parallel (θ_{12} large). The enhancement would seem to be limited, however, to the range over which $V(1,2)/kT$ is significant. If the range of the correlations proves to be much greater than that, we must surely infer that for nematic liquids (11)–(13) are insufficient.

3. Two reasons for doubt

It is my belief that the correlations of orientation represented by

$$\sigma_2(\boldsymbol{R}_{12}) - S_2^2$$

are in a typical nematic of far longer range than the possible range of $V(1,2)$ – that this quantity diminishes with distance, in fact, only as fast as $1/R$. If this assertion is correct, it not only casts doubt on the approximations on which Stecki and his coworkers have been forced to rely; it also implies that attempts to render the cluster model convincing, by steadily increasing the size of the clusters, are doomed to failure. A second reason for doubting the theories outlined in the previous section is that they include no allowance for the director fluctuations mentioned at the and of § 1. These two points will now be amplified.

Imagine a macroscopic cylinder whose sides have been 'rubbed' longitudinally to favour parallel alignment of the director in any nematic liquid brought into contact with them, and whose ends have been otherwise treated to favour perpendicular alignment. If immersed in a nematic with its axis parallel to the director, this cylinder need cause no distortion of the director field. If it is then tilted through an angle β_1, however, as shown in figure 2, the director will tilt too, by an angle $\beta(R)$, where R is measured from the centre of the cylinder. Now it is easily

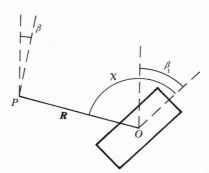

FIGURE 2. Rotation of the cylinder about O through an angle β_1 rotates the director at P through $\beta(R)$.

shown that when the three Frank constants are all equal, which is true of the model nematic investigated by Zannoni (1979) and others, though not of any real nematic yet identified, the configuration that minimizes the free energy stored in the nematic is such that $\beta(R)$ satisfies Laplace's equation
$$\nabla^2\beta = 0; \tag{14}$$
this is the form that the Euler–Lagrange equation takes for the problem in question. Hence we know that the equilibrium tilt can be expressed as a sum of spherical harmonic functions of the angle χ between R and the cylinder axis, i.e. that
$$\beta/\beta_1 = (a/R) + (b/R)^3\{\tfrac{1}{2}(3\cos^2\chi - 1)\} + O(1/R)^5. \tag{15}$$
At large distances only the first term in this series need be retained, and its coefficient a is clearly a length intermediate between half the length of the cylinder and its radius. Hence if θ is the angle between the axis of the cylinder and the z axis of a molecule at R, which is inclined at β_2 to the director at that point, we have
$$\langle P_2(\cos\theta)\rangle = \langle P_2(\cos\beta_2)\rangle P_2[\cos\{\beta_1 - (a\beta_1/R)\}]$$
$$= S_2\{P_2(\cos\beta_1) + 3(a\beta_1/R)\sin\beta_1\cos\beta_1 + O(a/R)^2\}. \tag{16}$$

Now imagine the cylinder to be, in fact, one of the molecules of which the nematic is composed. If correlations of orientation exist between nearest neighbours, any tilt of this central molecule must have some effect on the surrounding nematic, and we may reasonably hope to represent this effect by adjusting the length and radius of the cylinder and by treating everything outside the cylinder as a continuum in which (14) applies. The model suggests, if (16) is averaged over β_1, that at large distances
$$(\sigma_2(R) - S_2^2) = 3(a/R)S_2\langle\beta_1\sin\beta_1\cos\beta_1\rangle, \tag{17}$$

with a now a length of molecular dimensions. For well oriented nematics ($S_2 > 0.7$, say) we can expand thus:

$$3\langle\beta_1 \sin\beta_1 \cos\beta_1\rangle = 3\langle\sin^2\beta_1\rangle - \langle\sin^4\beta_1\rangle + \dots$$
$$= 2(1-S_2) - (56 - 80S_2 + 24S_4)/105 + \dots, \tag{18}$$

in the expectation that higher terms will be negligible. Hence we may estimate, using equation (26) below for S_4, that at large distances

$$\{\sigma_2(R) - S_2^2\} \approx 0.22(a/R) \quad \text{(for } S_2 = 0.867) \tag{19}$$
$$\approx 0.34(a/R) \quad \text{(for } S_2 = 0.753). \tag{20}$$

Evidence for the validity of these predictions is provided by figure 3, reproduced from Faber (1980). In this figure the crosses represent values of $\sigma_2(n)$ computed by Zannoni (1979) for the

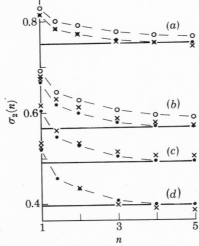

FIGURE 3. The short-range order parameter $\sigma_2(n)$ plotted against n, for a $10 \times 10 \times 10$ lattice of molecules subject to a nearest-neighbour interaction $-\epsilon P_2(\cos\theta_{ij})$: (a) $kT/\epsilon = 0.5$; (b) $kT/\epsilon = 0.8$; (c) $kT/\epsilon = 0.9$; (d) $kT/\epsilon = 1.0$. The filled circles are theoretical predictions, and the horizontal lines to which they converge for large n are predicted values of S_2^2 for the $10 \times 10 \times 10$ lattice at the temperature in question. The crosses show data generated by computer simulation. The open circles are calculated by using the same theory but for an infinite lattice. The values of kT/ϵ for which they apply are slightly different from 0.5 and 0.8, having been chosen to ensure that S_2^2 is the same as for the filled circles just below them. (Redrawn from Faber (1980).)

lattice model described above, plotted in effect against R; n is here the magnitude of R, measured in units of the lattice spacing, d. They are for four different values of (kT/ϵ), and the four horizontal lines represent the corresponding values of S_2^2. From these points alone one might conclude that the correlations of orientation, although appreciable up to $n = 3$, have virtually disappeared by the time $n = 4$ or 5. In this respect, however, results for a relatively small sample, only $10 \times 10 \times 10$ with periodic boundary conditions, give a misleading impression; had we relied on the computer-simulation results of Luckhurst & Romano (1980), who used a sample even smaller than Zannoni's (256 molecules, not confined to a lattice but kept apart by an isotropic Lennard–Jones potential, in a cubic cell with periodic boundary conditions), we would have concluded that the correlations of orientation were negligible beyond about two nearest-neighbour spacings. The open circles in the top two diagrams of figure 3 give a better picture of what $(\sigma_2 - S_2^2)$ should look like in an infinite lattice. Admittedly they are based upon a theory (see below) rather than on computer simulation, but the filled circles show that the theory is to be taken seriously; these filled circles were calculated in just the same way as the open circles but for a $10 \times 10 \times 10$ lattice, and they match Zannoni's crosses rather convincingly, considering that no adjustable parameter is involved. In the top two diagrams the open circles are converging for large n onto $(0.867)^2$ and $(0.753)^2$. The rate at which they do so seems to be nicely consistent with (19) and (20) respectively, with a taking the reasonable value of about $\frac{1}{3}d$, virtually independent of temperature.

That is the argument for believing that correlations of orientation in nematics are essentially of long range, varying as $1/R$. The argument for believing that director fluctuations are important in any realistic theory of nematics is equally straightforward. Light-scattering experiments indicate (Orsay Liquid Crystal Group 1969) that these fluctuations may be viewed as the super-position of a spectrum of periodic distortion modes, each of them excited, according to the equi-partition theorem, with mean square amplitude

$$\langle \psi^2 \rangle = kT/VKq^2 \tag{21}$$

for $k_{11} = k_{22} = k_{33} = K$; here ψ is the local angle of tilt of the director associated with a mode of wavevector q. Because excitation of any one mode may be shown to reduce the overall value of S_2 by a factor

$$\langle P_2(\cos \psi) \rangle = 1 - \tfrac{3}{2} \langle \psi^2 \rangle,$$

it follows that excitation of a spectrum of modes up to some cut-off wave vector q_c should reduce it from $S_2(0)$ to

$$S_2 = S_2(0) \exp\left(-\tfrac{3}{2} \sum_q \langle \psi^2 \rangle\right)$$

$$= S_2(0) \exp\left(-\frac{3kT}{(2\pi)^3} \int_0^{q^c} \frac{4\pi q^2 \mathrm{d}q}{K(q)\,q^2}\right); \tag{22}$$

a factor 2 in (22) takes care of the two polarizations that are allowed for each value of q. To illustrate the possible importance of the exponential we may take the model nematic discussed by Zannoni as an example. For this it may be shown (Faber 1980) that when $S_2 \approx 0.5$

$$K(q \to 0) \approx (kT/2d). \tag{23}$$

Hence, if we neglect the possible variation of $K(q)$ with q, we find

$$S_2 \approx S_2(0) \exp\left(-2d/\lambda_c\right), \tag{24}$$

where λ_c is the cut-off wavelength (i.e. $2\pi/q_c$).

Of course if one is to take into account only those modes for whose reality direct evidence is provided by light-scattering experiments, this means setting λ_c equal to about 200 nm, which is so much larger than the intermolecular spacing in a typical nematic as to make a quantity such as $2d/\lambda_c$ quite negligible. There is no real justification for cutting off the spectrum so soon, however. In Zannoni's lattice model, distortion modes of wavelength down to $2d$ are quite conceivable, and for such modes, indeed, $K(q)$ is likely to be rather smaller than for long wavelengths,† and the mean square amplitude correspondingly greater. Such considerations make the exponential correction factor in (24) look decidedly significant.

The contribution of director fluctuation to S_2 in nematics has been emphasized by a number of people besides myself, e.g. by Berreman (1975) and M. Warner (personal communication 1982). The long-range $(1/R)$ nature of orientation correlations is implicit in some results obtained by de Gennes (1969).

4. An alternative theory

The two criticisms of conventional theories that have been voiced in the previous section are arguably two sides of a single coin: the long-range tail in $(\sigma_2 - S_2^2)$ may be seen as an inevitable

† Equations quoted by Faber (1980) show that

$$K(q)/K(q \to 0) = 2(1 - \cos qd)/q^2 d^2.$$

122   T. E. FABER

consequence of the contribution that distortion modes of long wavelength make to the destruction of nematic order.

In my alternative theory (Faber 1977a, 1980), director fluctuations play the dominant role. It is assumed that but for these fluctuations the molecular alignment would be perfect, i.e. that $S_2(0)$ and likewise $S_4(0)$, etc., are unity. For the simple cubic lattice model of Zannoni the mode spectrum is cut off on the surfaces of a cubic Brillouin zone in q-space, i.e. on the planes $q_x = \pm \pi/d$, $q_y = \pm \pi/d$, $q_z = \pm \pi/d$. Otherwise the spectrum is cut off on a spherical surface of radius q_c, of sufficient volume to enclose $2N$ distinct modes, $2N$ being the number of orientational, as opposed to translational, degrees of freedom in a nematic system of N molecules. The theory leads to an expression for S_2 in terms of (kT/ϵ) and N for the lattice model or more generally in terms of the Frank constants k_{11}, k_{22} and k_{33}. It suggests a distribution function $f(\beta)$ such that

$$\ln S_n = \tfrac{1}{6}n(n+1)\ln S_2, \tag{25}$$

which means that

$$S_4 = S_2^{\frac{10}{3}}, \tag{26}$$

$$S_6 = S_2^7, \text{ etc.} \tag{27}$$

It enables quantities such as σ_2 and σ_4 to be worked out, and for the simple cubic lattice model (at any rate for $R = d$ and in the limit $N \to \infty$) it predicts a relation between these quantities, which may be expressed to a high degree of accuracy by the equation (Faber 1981)

$$\sigma_4 = \sigma_2^{\frac{10}{3}}. \tag{28}$$

It has been used to discuss the order-dependence of the Frank constants (Faber 1977b, 1981), with results that differ significantly from those of mean field theories. Finally, it provides a way of calculating time-dependent correlation functions of relevance to the theories of nuclear magnetic relaxation (Faber 1977c) and dielectric dispersion in nematics; direct information about the behaviour of these correlation functions in model nematics is beginning to be available from computer-simulation studies (Zannoni & Guerra 1981).

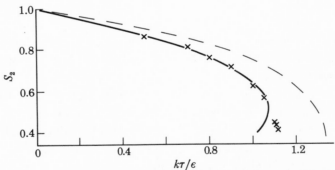

FIGURE 4. Dependence of S_2 on (kT/ϵ) for the $10 \times 10 \times 10$ lattice of figure 3. The full curve represents the theoretical prediction and the crosses show data generated by computer simulation. (The broken curve is predicted by mean field theory.)

The success of this so-called 'continuum theory of nematic disorder' in matching, without any adjustable parameter, Zannoni's results for $\sigma_2(n)$ against n in a $10 \times 10 \times 10$ sample has been illustrated in figure 2 above. Figure 4 shows that it can also match, again without any adjustable parameter, his results for S_2 against (kT/ϵ) in the same sample, except in the region near $T = T_c$, where S_2 falls below, say, 0.5.

Figure 5 shows how S_4 varies with S_2, not just for the model studied by Zannoni (1979) and Zannoni & Guerra (1981) but for two quite different models – smaller in size, involving different intermolecular potentials, and with no constraining lattice – on which Tsykalo & Bagmet (1978a, b) have performed computer-simulation experiments. The figure includes points showing how, in Zannoni's model, $\sigma_4(d)$ varies with $\sigma_2(d)$. The data seem to confirm that there is a more or less unique, model-independent, relation between S_4 and S_2 and that the relation between $\sigma_4(d)$ and $\sigma_2(d)$ is very similar. At low temperatures it is accurately described by (20) or (28) as the case may be, but systematic deviations become apparent when S_2 or σ_2 fall below, say, 0.6.

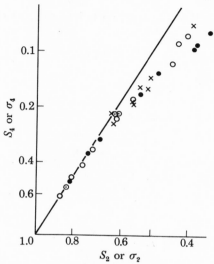

FIGURE 5. Relation between S_4 and S_2, or between $\sigma_4(d)$ and $\sigma_2(d)$, on a logarithmic plot. The points represent data generated by computer simulation by using three distinct models: ○, Zannoni (1979) (S_4 against S_2); ⊙, Zannoni & Guerra (1981) (S_4 against S_2); ×, Tsykalo & Bagmet (1978a, b) (S_4 against S_2); ●, Zannoni (1979) (σ_4 against S_2). The straight line represents the predicted $\frac{10}{3}$ power law.

Zannoni & Guerra (1981) have provided two values for S_6 in the $10 \times 10 \times 10$ lattice model that may be compared with predictions based on (27). Where S_2 is 0.833 the predicted value is 0.278: the computed value is 0.298 ($\pm$?). Where S_2 is 0.620 ($\pm$ 0.005) the predicted value is 0.035 ($\pm$ 0.003): the computed value is 0.05 ($\pm$?). The agreement is not unsatisfactory.

By and large, therefore, the results of computer-simulation studies confirm the reliability of the continuum approach at low temperatures but show that it cannot be trusted to describe the transition into the isotropic phase, at which S_2 is often as low as 0.35. These conclusions are scarcely a surprise. Since the theory takes the perfectly aligned state as its starting point – whereas theories like that of Stecki & Kloczkowski (1979) tend to start from the isotropic phase – it is bound to work best at low temperatures. Since the r.m.s. amplitude of each distortion mode depends (see (21)) on the magnitude of K, and since the magnitude of K depends upon the degree to which all the other modes are excited, an element of approximation enters the theory at an early stage, where the modes are assumed to be independent of one another and therefore excited with random phase. It is presumably this approximation that begins to fail as T_c is approached.

To test the theory against the behaviour of real nematics one needs first to measure, with considerable precision, the three Frank constants k_{11}, k_{22} and k_{33} for small q and then to evaluate the appropriate average of these to use in expressions such as (21) and (22) above. An equation defining the average has been given by Faber (1977a), though he has emphasized that it may give too much weight to the 'bend' constant k_{33}, and since this does not necessarily vary with temperature in quite the same way as k_{11} and k_{22} the use of that equation may introduce some error. Then if reliable values of S_2 are available one may plot $\ln S_2$ against $(T\rho^{\frac{1}{3}}/K)$, where ρ is density. A straight line is to be expected, at any rate for temperatures such that $S_2 > 0.5$ or 0.6. The intercept on the $\ln S_2$ axis should tell us $S_2(0)$, and this is expected to be unity. The slope should be given (Bunning *et al.* 1981) by

$$-\ln S_2/(T\rho^{\frac{1}{3}}/K) \approx 0.76k(N/M)^{\frac{1}{3}}. \tag{29}$$

The numerical coefficient in (29) depends upon how $K(q)$ varies with q; the figure 0.76 is based upon a detailed analysis of the lattice model discussed above, and for real nematics it may be in error by 20 % or so.

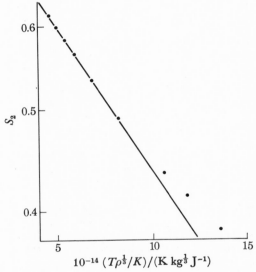

FIGURE 6. Test of continuum theory for nematic 5CB. The scale for S_2 is logarithmic.

Figure 6 shows such a plot for the nematic for which the most reliable information is currently available, 4'-*n*-pentyl-4-cyanobiphenyl or 5CB. It is an updated version of a figure plotted already by Bunning *et al.* (1981), using nine values of S_2 that Emsley *et al.* (1981) have derived from the proton n.m.r. spectrum of a 5CB specimen in which all but the four protons on one of the phenyl rings have been replaced by deuterons, and using our own values of the magnetic anisotropy $\Delta\chi$ to calculate k_{11}, k_{22} and k_{33}, because we now have reason (Faber *et al.* 1983) to prefer these to the Groningen data on which Bunning *et al.* relied. A straight line can indeed be fitted to the points for which $S_2 > 0.5$. Its intercept corresponds to $S_2 = 0.82$ rather than unity, however, and its slope is 0.63×10^{-15} J kg$^{-\frac{1}{3}}$ K^{-1}, whereas the slope predicted by (29) is 1.4×10^{-15} J kg$^{-\frac{1}{3}}$ K^{-1}. We have enough information available about the properties of other cyanobiphenyl derivatives to know that they would all show similar discrepancies.

All cyanobiphenyl derivatives are strongly polar, of course, and there is evidence from X-ray diffraction (Leadbetter *et al.* 1975) and from dielectric permittivity measurements (Chandrasekhar 1977) that neighbouring molecules associate on this account. One way to explain why the slope of the straight line in figure 6 is about half the expected value is to postulate that the molecules form strongly associated pairs, because pairing would halve the number of orientational degrees of freedom and therefore halve the number of distortion modes contributing to $\Sigma_q\langle\psi^2\rangle$ (see(22)).

Pairing would explain the anomalous intercept too, if it could be shown that the two members of a pair are not collinear but lie with their z axes inclined to each other at an angle of about $40°$. It is probable, however, that other complicating factors are involved, including the secondary ordering described by $(S_{xx} - S_{yy})$ and the non-rigid character of that part of each molecule that consists of an alkyl chain. Ideally, the theory should be tested against the behaviour of a non-polar nematic without a flexible end chain, for which $(S_{xx} - S_{yy})$ is known to be zero.

5. CONCLUSIONS

The idea of pairing between molecules of 5CB brings us back to the cluster model of §2 and may indicate the way forward to a more realistic theory of nematics than any yet discussed. If we could find some way to deal exactly with the correlations of orientation and position to be expected within a cluster of, say, a dozen molecules, we could hope to describe correlations between clusters by using the theory outlined in §4. Complete understanding of the situation near T_c, however, where S_2 is neither small nor close to unity, may remain elusive for many years yet.

REFERENCES

Berreman, D. W. 1975 *J. chem. Phys.* **62**, 776–778.

Bunning, J. D., Faber, T. E. & Sherrell, P. L. 1981 *J. Phys., Paris* **42**, 1175–1182.

Chandrasekhar, S. 1977 *Liquid crystals.* Cambridge University Press.

Chandrasekhar, S. & Madhusudana, N. V. 1971 *Acta crystallogr.* A **27**, 303–313.

Chandrasekhar, S. & Madhusudana, N. V. 1973 *Molec. Cryst. liq. Cryst.* **24**, 179–186.

de Gennes, P. G. 1969 *Molec. Cryst. liq. Cryst.* **7**, 325–345.

de Gennes, P. G. 1974 *The physics of liquid crystals.* Oxford: Clarendon Press.

Emsley, J. W., Luckhurst, G. R. & Stockley, C. P. 1981 *Molec. Phys.* **44**, 565–580.

Faber, T. E. 1977a *Proc. R. Soc. Lond.* A **353**, 247–259.

Faber, T. E. 1977b *Proc. R. Soc. Lond.* A **353**, 261–275.

Faber, T. E. 1977c *Proc. R. Soc. Lond.* A **353**, 277–288.

Faber, T. E. 1980 *Proc. R. Soc. Lond.* A **370**, 509–521.

Faber, T. E. 1981 *Proc. R. Soc. Lond.* A **375**, 579–597.

Faber, T. E., Bunning, J. D. & Crellin, D. A. 1983 (In preparation.)

Humphries, R. L., James, P. G. & Luckhurst, G. R. 1972 *J. chem. Soc. Faraday Trans. II* **68**, 1031–1044.

Leadbetter, A. J., Richardson, R. M. & Colling, C. N. 1975 *J. Phys. Colloq., Paris* **36**, C1-37–C1-43.

Luckhurst, G. R. & Romano, S. 1980 *Proc. R. Soc. Lond.* A **373**, 111–130.

Luckhurst, G. R. & Zannoni, C. 1977 *Nature, Lond.* **267**, 412–414.

Luckhurst, G. R., Zannoni, C., Nordio, P. L. & Segre, U. 1975 *Molec. Phys.* **30**, 1345–1358.

Maier, W. & Saupe, A. 1958 *Z. Naturf.* **13a**, 564–566.

Maier, W. & Saupe, A. 1959 *Z. Naturf.* **14a**, 882–889.

Maier, W. & Saupe, A. 1960 *Z. Naturf.* **15a**, 287–292.

Onsager, L. 1949 *Ann. N.Y. Acad. Sci.* **51**, 627–659.

Orsay Liquid Crystal Group 1969 *J. chem. Phys.* **51**, 816–822.

Ruijgrok, T. W. & Sokalski, K. 1982 *Physica* A **111**, 45–64.

Shen, J., Lin, L., Yu, L. & Woo, C. 1981 *Molec. Cryst. liq. Cryst.* **70**, 301–313.

Sheng, P. & Wojtowicz, P. J. 1976 *Phys. Rev.* A **14**, 1883–1894.

Stecki, J. & Kloczkowski, A. 1979 *J. Phys. Colloq., Paris* **40**, C3-360–C3-362.

Tsykalo, A. L. & Bagmet, A. D. 1978a *Soviet Phys. solid State* **20**, 762–766.

Tsykalo, A. L. & Bagmet, A. D. 1978b *Molec. Cryst. liq. Cryst.* **46**, 111–119.

Wagner, W. 1981 *Molec. Cryst. liq. Cryst.* **75**, 169–177.

Zannoni, C. 1979 In *The molecular physics of liquid crystals* (ed. G. R. Luckhurst & G. W. Gray), ch. 2, pp. 191–219. New York: Academic Press.

Zannoni, C. & Guerra, M. 1981 *Molec. Phys.* **44**, 849–869.

Discussion

Lin Lei (*Institute of Physics, Chinese Academy of Sciences, Beijing, China*). Does Dr Faber's theory produce $\langle P_4 \rangle < 0$, which are observed in some experiments on nematics?

T. E. Faber. No, unless one is prepared to postulate that the quantity that I have called $S_4(0)$ is negative. The sort of model I have suggested in my paper to explain values of $S_2(0)$ as low as 0.82 could in principle be used to explain negative values of $S_4(0)$.

L. G. P. Dalmolen (*Solid State Physics Laboratory, University of Groningen, The Netherlands*). Recently, negative values of the order parameter $\langle P_4 \rangle$ of different compounds, measured by polarized Raman scattering, have been reported in the literature. Theoreticians seem to be worried about that because mean field and other theories cannot predict such negative results for $\langle P_4 \rangle$. We (Dalmolen & de Jeu, *J. chem. Phys.* (in the press)) measured the order parameters of various compounds, among which some were reported to have a negative $\langle P_4 \rangle$, with polarized Raman scattering. In all cases either a mean field behaviour of $\langle P_4 \rangle$ was observed or lower values between 0.05 and 0.15, but not negative. In our opinion the negative values of $\langle P_4 \rangle$ reported can be ascribed to either inadequate correction of the scattering of the sample or improper calibration. The latter point is evident from large discrepancies in the values of R_{iso}, the depolarization in the isotropic phase. We therefore feel that the problem of the negative values of $\langle P_4 \rangle$ is just an experimental and not a theoretical one.

Sir Charles Frank, F.R.S. (*The University, Bristol*). I see no reason to doubt the accuracy of some experimental observations just because they yield a negative value for $\langle P_4 \rangle$. To doubt it because in some other cases $\langle P_4 \rangle$ is positive is like saying 'I have looked at 20 mountains which didn't have craters at the top: therefore there are no volcanoes'. I think so long as one makes theories with cylindrical molecules one will predict a positive $\langle P_4 \rangle$, but if one makes the theory for molecules shaped like dog-bones, negative $\langle P_4 \rangle$ is a likely outcome: and on the whole molecules are rather more like dog-bones than they are like pencils.

Phil. Trans. R. Soc. Lond. A **309**, 127–144 (1983)
Printed in Great Britain

The van der Waals theory of nematic liquids

By Martha A. Cotter

Department of Chemistry, Rutgers – The State University,
New Brunswick, New Jersey 08540, U.S.A.

This paper describes the van der Waals theory of nematic liquids, an approximate molecular theory in which very short-range intermolecular repulsions are approximated by hard-rod exclusions, and somewhat longer-ranged intermolecular attractions are subject to a self-consistent mean-field treatment. The rationale, underlying assumptions, idealizations and approximations of the theory are presented in detail and the numerical results so far reported are summarized, together with the results of extensive new calculations, which provide a quite accurate test of the theory in its present state. Finally, the current status of the theory, its relative strengths and weaknesses, and the prospects for extending and improving it are discussed.

1. Introduction

A nematic mesophase is a very complex condensed system for which relatively little is known about the precise form of the intermolecular pair potential. It is therefore clearly impossible to devise a realistic molecular theory of nematogens. Instead, any tractable molecular theory of these systems must incorporate a number of rather severe idealizations or approximations. What sort of idealized model or statistical mechanical approximations to invoke depends on the purpose for which the theory is intended.

At present, the most important aspects of nematic mesomorphism which need to be addressed by molecular theories are (1) the nature of the nematic–isotropic (N–I) and the nematic–smectic A (S–N) phase transitions and (2) the relation between molecular structure and the stability and properties of nematic mesophases. For investigating critical phenomena in the vicinity of the N–I or S–N transition temperatures, the molecular details of the model system are presumably not important as long as the model Hamiltonian has the correct symmetry. On the other hand, it is important that the statistical mechanics be done as accurately as possible. In this case, therefore, one can profitably use a very highly idealized model system such as an array of point particles on a three-dimensional lattice interacting through a nearest-neighbour pair potential $-JP_2(\cos\theta_{ij})$, where J is a positive constant, P_2 is the Legendre polynomial of order 2, and θ_{ij} is the angle between the unit vectors a_i and a_j, which characterize the 'orientations' of particles i and j respectively. On the other hand, such a simple model is clearly inadequate for investigating the relation between molecular structure and the stability and properties of nematic phases. For this purpose, a more 'chemical' model is needed, i.e. a model for which it is a relatively straightforward task to build in what is known about the size, shape, flexibility, polarity, polarizability, etc. of a particular nematogen. The van der Waals theory of nematic liquids is intended to be such a 'chemist's' theory. Conversely, it is not designed for the study of critical phenomena in nematogenic systems.

In addition to being a 'chemist's' approach, the van der Waals theory may be viewed as a hybrid of what were once the two rival theoretical approaches to nematogens: (1) the Maier–

Saupe theory (Maier & Saupe 1959, 1960) and its extensions and modifications (see Luckhurst (1979) for a review of Maier–Saupe-type theories) and (2) the so-called 'hard-rod' theories of nematics (see Cotter (1979a) for a review of hard-rod theories). In the former it was assumed that nematic order results primarily from the anisotropy of the intermolecular attractions, particularly London dispersion forces, and these attractions were treated in the mean-field approximation. In the latter it was assumed that nematic order results primarily from excluded volume effects, and model systems of hard (i.e. infinitely impenetrable) rod-like objects were therefore used. In the van der Waals theory, the model molecules have hard cores as in hard-rod theories, but they also exhibit anisotropic intermolecular attractions, and the latter are treated in the mean-field approximation. The need for hybrid theories incorporating both anisotropic intermolecular repulsions *and* attractions was made clear over the years by a variety of experimental results.

Finally, the van der Waals theory of nematogens is a historical and logical descendant of the van der Waals theory of simple liquids, which employed a model system of hard spheres in a spatially uniform mean-field arising from intermolecular attractions. This approach was used quite successfully by Longuet-Higgins & Widom (1964) to calculate the thermodynamic properties of liquid argon near its triple point. This success leads one to hope that a similar approach with the use of rod-like molecular hard cores and an orientation-dependent mean field can be used successfully to treat nematic liquids.

The basic assumptions, idealizations and approximations of the van der Waals theory are described in §2 of this paper, and the numerical results so far obtained are summarized in §3. Finally, the current status of the theory and the prospects for improving and extending it are discussed in §4.

2. BASIC ASSUMPTIONS, IDEALIZATIONS AND APPROXIMATIONS

The van der Waals theory of nematic liquids, like the van der Waals theory of simple liquids, is based on three underlying assumptions: (1) that the structure of a liquid far from the gas–liquid critical point is largely determined by very short-ranged intermolecular repulsions; (2) that these repulsions may satisfactorily be approximated by hard-core exclusions; and (3) that the primary role of the somewhat longer-ranged intermolecular attractions (dispersion forces, etc.) is – to a first approximation – to provide a negative, spatially uniform mean field in which the molecules move. For simple liquids, assumptions (1) and (2) are clearly reasonable, since the radial distribution function for liquid argon is barely distinguishable from that of a fluid of hard spheres of appropriate radius. Moreover, the suitability of assumption (3) can be inferred from the success of Longuet-Higgins & Widom's calculations for liquid argon. Since nematic liquids differ from isotropic liquids only in their exhibition of long-range orientational order, it would seem that these underlying assumptions should also be reasonable for nematogens. In practice, of course, applying the van der Waals approach to a nematic liquid is more complicated and risky than applying it to a simple liquid owing to the difficulty in choosing an appropriate shape for the molecular hard cores for a particular nematogen and the necessity to consider molecular orientations as well as positions.

Van der Waals theories of nematics based on rod-like molecular hard cores and an orientation-dependent mean field have been devised by Cotter (1977a) and Gelbart & Baron (1977). In the former instance, an empirical mean-field pseudo-potential of the Maier–Saupe type was used. In the generalized van der Waals theory of Gelbart & Baron, on the other hand, the mean-field pseudo-potential was calculated from an assumed intermolecular pair potential in a self-consistent manner. A more compact and straightforward derivation of the generalized van der Waals theory was later presented by Cotter (1977 b).

In this paper, the statistical mechanical derivation of the van der Waals theory will not be presented, because it is available elsewhere. (The interested reader is referred to the original papers of Gelbart & Baron (1977) and Cotter (1977 a, b) or to the review chapter by Cotter (1979 b).) Instead, the idealizations and approximations of the theory will be described in detail and the final results of the statistical mechanical manipulations will then be presented.

Without further introduction, therefore, the idealizations and approximations of the van der Waals theory of nematogens are described below.

1. Molecular flexibility is neglected; i.e. the dependence of the potential energy of the system on the intramolecular conformational coordinates of the molecules is neglected and a nematogen is modelled as an inflexible object with some average or effective shape. The potential energy of interaction between two molecules i and j is thus assumed to depend only on the orientations Ω_i and Ω_j of the two molecules and on the vector $\mathbf{r}_{ij}$ between their centres. Specifically, the N-body potential energy is assumed to be additive pairwise and the intermolecular pair potential v is assumed to have the form

$$v(\mathbf{r}_{ij}, \Omega_i, \Omega_j) = v_{\text{rep}}(\mathbf{r}_{ij}, \Omega_i, \Omega_j) + v_{\text{att}}(\mathbf{r}_{ij}, \Omega_i, \Omega_j), \tag{1}$$

where v_{rep} is a very short-range repulsive potential and v_{att} is a somewhat longer-ranged attractive potential.

2. The very short-range intermolecular repulsions are approximated by hard-core exclusions; i.e. it is assumed that

$$v_{\text{rep}}(\mathbf{r}_{ij}, \Omega_i, \Omega_j) = v^*(\mathbf{r}_{ij}, \Omega_i, \Omega_j)$$

$$= \begin{cases} \infty & \text{if the hard cores of } i \text{ and } j \text{ would overlap;} \\ 0 & \text{otherwise.} \end{cases}$$

3. The rod-like molecular hard cores are taken to have cylindrical symmetry. Ω thus represents a set of two angles, the polar angle θ and the azimuthal angle ϕ, which specify the orientation of a cylindrical molecular axis with respect to a space-fixed coordinate system.

4. Intermolecular attractions are treated via the mean-field approximation; i.e. the attractions between a molecule with orientation Ω and all the other molecules are approximated collectively by the interaction of the given molecule with a spatially uniform mean field described by the pseudo-potential $\overline{\psi}(\Omega, \rho)$. As noted above, two different approaches have been used to evaluate the pseudo-potential $\overline{\psi}(\Omega, \rho)$.

(a) In the empirical approach first used by Cotter (1977 a), the pseudo-potential was assumed to have the simple functional form

$$\overline{\psi}(\Omega, \rho) = -v_0 \rho - v_2 \rho \eta_2 P_2(\cos\theta), \tag{2}$$

where ρ is the number density, v_0 and v_2 are positive constants, θ is the angle between the cylindrical axis of the molecular hard core and the nematic director (taken to be parallel to the z axis of the space-fixed coordinate system), P_2 is the Legendre polynomial of order 2, and η_2, the traditional nematic order parameter, is the average value of $P_2(\cos\theta)$. From a statistical mechanical point of view,

$$\eta_2 = \int f(\Omega) P_2(\cos\theta) \, \mathrm{d}\Omega, \tag{3}$$

where $f(\Omega)$ is the normalized one-body orientational distribution function formally defined by the statement that $f(\Omega)\,\mathrm{d}\Omega$ is the fraction of molecules with orientations between Ω and $\Omega + \mathrm{d}\Omega$. In this approach, v_0 and v_2 are treated as adjustable parameters and no attempt is made to calculate the ratio v_2/v_0 from an assumed intermolecular pair potential. Finally, it should be noted that the linear dependence of $\overline{\psi}$ on ρ is required for statistical mechanical self-consistency (Cotter 1977c).

(b) In the generalized van der Waals theory of Gelbart & Baron (1977), the mean-field pseudo-potential $\overline{\psi}$ is related to the intermolecular pair potential by the equation

$$\overline{\psi}(\Omega, \rho) = \rho \int \mathrm{d}\Omega' f(\Omega') \int \mathrm{e}^{-\beta v^*(r, \Omega, \Omega')} v_{\mathrm{att}}(r, \Omega, \Omega')\,\mathrm{d}r$$

$$\equiv \rho \int \mathrm{d}\Omega' f(\Omega')\,I(\Omega, \Omega'), \tag{4}$$

where $\beta = 1/kT$ (k is Boltzmann's constant) and r is the vector connecting the centres of two molecules with orientations Ω and Ω' respectively. Since v^* can have only two values, zero or infinity, the Boltzmann factor $\mathrm{e}^{-\beta v^*}$ serves only to exclude from the range of integration over r all values of r that correspond to overlap of the hard cores of the two given molecules. It is therefore clear that in evaluating $\overline{\psi}$, all short-range correlations between pairs of molecules are neglected *except* for the requirement that their hard cores not overlap.

5. The contribution of the molecular hard cores to the free energy of the system is calculated by using scaled particle theory, an approximate statistical mechanical theory of fluids of hard particles. (See below for an outline of this approach.)

Having made assumptions 1–4 above, it can be shown that the Helmholtz free energy functional for the model system is given by

$$N^{-1}A\{f(\Omega)\} = N^{-1}A^*\{f(\Omega)\} + \tfrac{1}{2}\int f(\Omega)\,\overline{\psi}(\Omega)\,\mathrm{d}\Omega, \tag{5}$$

where $A^*\{f(\Omega)\}$ is the free energy functional for a system of hard rods constrained to have orientational distribution function $f(\Omega)$. The most convenient cylindrically symmetric shape to choose for the molecular hard cores is a spherocylinder (i.e. a right circular cylinder capped on each end by a hemisphere of the same radius) because the mutual exclusion volume of two spherocylinders has a particularly simple dependence on their orientations. One is then confronted by the task of evaluating the Helmholtz free energy of a fluid of hard spherocylinders. This clearly cannot be done exactly. Furthermore, no computer simulation of a system of three-dimensional hard rods, spherocylindrical or otherwise, has succeeded in observing a nematic phase. (There have been successful Monte Carlo simulations for isotropic fluids of hard spherocylinders (see below), but at the high densities apparently required to observe the nematic phase, convergence of the Monte Carlo computations becomes impracticably slow.) Thus some approximate statistical mechanical method must be used to evaluate $A^*\{f(\Omega)\}$. As noted above, in the work so far on the van der Waals theory, the approach used has almost always been scaled particle theory.

Scaled particle theory is an approximate statistical mechanical theory of fluids of hard particles first developed by Reiss *et al.* (1959) to treat systems of hard spheres. For fluids of both hard spheres and hard discs, its predictions agree very well with the results of Monte Carlo and molecular dynamics computer simulations. The theory has been extended to treat systems of hard spherocylinders by a number of authors (Cotter & Martire 1970a, b; Lasher 1970; Timling 1974;

Cotter 1974, 1977 a). Although the precise mathematical expressions derived for $A*\{f(\Omega)\}$ differ somewhat among these various versions of the theory, the numerical results obtained differ very little. In what follows, the version of Cotter (1977 a) will be discussed for concreteness.

In this version, the central quantity of scaled particle theory for a fluid of hard spherocylinders of radius a and cylindrical length l is the work function $W(\alpha, \lambda, \Omega)$, which is defined as the reversible work necessary to add a scaled spherocylinder of radius αa, cylindrical length λl, and orientation Ω to the system at an arbitrary fixed point. This quantity is related to the chemical potential of the system by the exact relation

$$\mu - \mu_{\text{ideal}} = kT \int f(\Omega) \ln [4\pi f(\Omega)] \, d\Omega + \int f(\Omega) \, W(1,1,\rho) \, d\Omega. \qquad (6)$$

Although an exact expression for $W(\alpha, \lambda, \Omega)$, valid for all α and λ, cannot be derived, the exact limiting behaviour of W is known for very small and very large scaled spherocylinders, i.e.

$$\lim_{\substack{\alpha \to 0 \\ \lambda \to 0}} W(\alpha, \lambda, \Omega) = -kT \ln (1 - v_0 \rho),$$

where $v_0 = \pi a^2 l + \frac{4}{3}\pi a^3$ is the volume of a hard spherocylinder, and

$$\lim_{\substack{\alpha \to \infty \\ \lambda \to \infty}} W(\alpha, \lambda, \Omega) = P\{\pi(\alpha a)^2 \lambda l + \frac{4}{3}\pi(\alpha a)^3\},$$

where P is the external pressure. (The latter limit simply recognizes the fact that when the scaled spherocylinder becomes very large, W must equal the reversible thermodynamic work of building a macroscopic cavity in the fluid.) Between these two limits, one approximates $W(\alpha, \lambda, \Omega)$ by using the interpolation formula

$$W(\alpha, \lambda, \Omega) = -\ln (1 - v_0 \rho) + C_{10}(\Omega, a, l, \rho) \, \alpha + C_{01}(\Omega, a, l, \rho) \, \lambda$$
$$+ C_{11}(\Omega, a, l, \rho) \, \alpha\lambda + C_{20}(\Omega, a, l, \rho) \, \alpha^2 + (\pi a^2 l \beta P) \, \alpha^2 \lambda + (\tfrac{4}{3}\pi a^3 \beta P) \, \alpha^3, \qquad (7)$$

where $C_{ij} = (i!j!)^{-1} (\partial^{(i+j)} \beta W/\partial \alpha^i \partial \lambda^j)_{\alpha=0, \lambda=0}$. (The derivatives needed to calculate C_{10}, C_{01}, C_{11} and C_{20} can be evaluated exactly.) The equilibrium thermodynamic properties of the system can then be obtained from $\overline{W}(1,1) = \int f(\Omega) \, W(1,1,\Omega) \, d\Omega$. When the equation of state obtained in this manner for an isotropic fluid of hard spherocylinders is compared with that calculated from computer simulations (Vieillard-Baron 1974; Monson & Rigby 1978), it is found that the agreement is quite good at low to moderate densities, but that the scaled particle predictions for $Pv_0/(kT)$ become significantly and increasingly too large at reduced densities $(v_0 \rho)$ of order 0.50. For example, for the highest reduced densities at which the computer simulations were carried out for spherocylinders with length : width ratio $x = 3$ (Vieillard-Baron 1974), namely $v_0 \rho = 0.50$ and $v_0 \rho = 0.54$, the scaled particle results are too high by 17.3 % and 23.5 % respectively. For spherocylinders with $x = 2$, on the other hand, the highest reduced density considered in the computer simulations (Monson & Rigby 1978) was $v_0 \rho = 0.5096$, and the scaled particle prediction is too high by 7.9 % at that point. These discrepancies between theory and computer 'experiment' for the isotropic phase at high densities suggest that the use of scaled particle theory may be a significant source of error in the van der Waals theory. Finding a more accurate but tractable method of evaluating $A*\{f(\Omega)\}$ is, however, a quite difficult task.

When the expression for $A*\{f(\Omega)\}$ derived from this version of scaled particle theory is inserted in (5), the final result for the van der Waals free energy functional is

$$\frac{A*\{f(\Omega)\}}{NkT} = \langle \ln \{4\pi f(\Omega)\} \rangle + \ln \{\rho/(1 - v_0 \rho)\} + \frac{(4 + q - \frac{1}{2}q^2) \, (v_0 \rho)^2}{3(1 - v_0 \rho)^2}$$
$$+ \tfrac{1}{2}\lambda(a, l, \rho) \langle\langle |\sin \gamma(\Omega, \Omega')| \rangle\rangle + (2kT)^{-1} \int f(\Omega) \, \overline{\psi}(\Omega) \, d\Omega, \qquad (8)$$

[61]

where
$$q = \frac{4\pi a^3}{3v_0}, \quad r = \frac{al^2}{v_0}, \quad \lambda = \frac{4rv_0\rho\{1 - \frac{1}{3}(1-q)\,v_0\rho\}}{(1 - v_0\rho)^2},$$

$$\langle \ln\{4\pi f(\Omega)\}\rangle = \int f(\Omega)\ln\{4\pi f(\Omega)\}\,\mathrm{d}\Omega,$$

$$\langle\langle|\sin\gamma|\rangle\rangle = \iint |\sin\gamma(\Omega, \Omega')|f(\Omega)f(\Omega')\,\mathrm{d}\Omega\,\mathrm{d}\Omega', \tag{9}$$

and $\gamma(\Omega, \Omega')$ is the angle between directions Ω and Ω'. The orientational distribution function, $f(\Omega)$, is obtained by minimization of the free energy functional. The result is the nonlinear integral equation

$$f(\Omega) = C\exp\left[-\left\{\beta\overline{\psi}(\Omega, \rho) + \lambda(a, l, \rho)\int \mathrm{d}\Omega'f(\Omega')\,|\sin\gamma(\Omega, \Omega')|\right\}\right], \tag{10}$$

where C is a normalization factor.

3. Summary of results

(a) Results obtained by using an empirical pseudo-potential

Cotter (1977a) carried out extensive calculations for a model system with hard-core length: width ratio $x = l/2a + 1 = 3$ and compared the results with experimental data for the much-studied nematogen p-azoxyanisole (PAA). The hard-core volume v_0 was taken to be $0.230\,\mathrm{nm}^3$ and the mean field pseudo-potential was assumed to have the form (2) with $v_0/(v_0 k) = 25\,000\,\mathrm{K}$ and $v_2/(v_0 k) = 2000\,\mathrm{K}$. The choices for x and v_0 were suggested by Vieillard-Baron (1974), based on estimates from tabulated bond lengths and van der Waals radii. On the other hand, the parameters v_0 and v_2 were chosen to reproduce the experimental values of the transition temperature T_{NI} and the quantity $\tau = -\rho(\partial\eta_2/\partial\rho)_T/T(\partial\eta_2/\partial T)_\rho$ at T_{NI} for PAA. (The latter is a measure of the relative sensitivity of the order parameter η_2 to changes in ρ against changes in T.) In addition to those described above, one further approximation was introduced, namely

$$\langle\langle|\sin\gamma(\Omega, \Omega')|\rangle\rangle \approx \tfrac{1}{4}\pi - \tfrac{5}{32}\pi\eta_2^2, \tag{11}$$

which is obtained by expanding $|\sin\gamma|$ in even-order Legendre polynomials $P_{2n}(\cos\gamma)$, averaging term by term, and then truncating the resulting series after its second term. This should be a reasonably good approximation for relatively small values of η_2, but not for values of η_2 close to unity. For hard spherocylinders with $x = 3$, it lowers the predicted values of the order parameter η_2 by roughly 25 % but changes only slightly the densities of the coexisting phases at the N–I transition. It greatly simplifies the integral equation (10) for $f(\Omega)$, which becomes

$$f(\Omega) = C\exp\{\Lambda(\rho, T)\,\eta_2 P_2(\cos\theta)\} = \frac{\exp\{\Lambda(\rho, T)\,\eta_2 P_2(\cos\theta)\}}{2\pi\displaystyle\int_0^\pi \exp\{\Lambda(\rho, T)\,\eta_2 P_2(\cos\theta)\}\sin\theta\,\mathrm{d}\theta}, \tag{12}$$

where
$$\Lambda(\rho, T) = \tfrac{5}{32}\pi\lambda(a, l, \rho) + v_2/kT. \tag{13}$$

Determination of $f(\Omega)$ thus reduces to solving iteratively for η_2 by using the self-consistency condition

$$\eta_2 = \int f(\Omega)\,P_2(\cos\theta)\,\mathrm{d}\Omega = \frac{\displaystyle\int_0^\pi P_2(\cos\theta)\exp[\Lambda(\rho, T)\,\eta_2 P_2(\cos\theta)]\sin\theta\,\mathrm{d}\theta}{\displaystyle\int_0^\pi \exp[\Lambda(\rho, T)\,\eta_2 P_2(\cos\theta)]\sin\theta\,\mathrm{d}\theta}. \tag{14}$$

To summarize the results of these calculations, it can be said that quite satisfactory qualitative but not quantitative agreement with experiment was obtained. The theory predicts η_2–T curves (at constant P or constant ρ) of roughly the correct shape, nearly linear plots of $\ln T$ against $\ln \rho$ at constant η_2 with slopes close to 4 (in agreement with the data of McColl & Shih (1972)), increases in T_{NI} with increasing P of the correct order of magnitude, and large pre-transitional increases in the compressibility, expansivity and specific heat as T_{NI} is approached from below. (For more details, see Cotter (1977a).) On the other hand, the N–I phase transition is predicted to be much too strongly first-order, as can be seen from the first two columns of table 1.

TABLE 1. VALUES OF VARIOUS QUANTITIES AT THE N–I PHASE TRANSITION AS PREDICTED BY THE VAN DER WAALS THEORY WITH THE EMPIRICAL MEAN FIELD POTENTIAL

$$\bar{\psi}(\Omega, \rho) = -v_0 \rho - v_2 \rho \eta_2 P_2(\cos \theta)$$

($P = 1$ atm ($ca.$ 10^5 Pa) in all cases.)

quantity	experimental data for PAA†	theoretical predictions		
		$x = 3$‡	$x = 1.75$‡	$x = 1.75$§
$\{v_0/(v_0 k)\}/10^4$ K	—	2.50	5.42	5.38
v_2/v_0	—	0.0800	0.0326	0.0324
$T_{NI}/$K	409	410	409	409
$v_0 \rho_{\text{nem}}$	0.62	0.445	0.62	0.62
$\Delta\rho/\rho_{\text{nem}}$	0.0035	0.040	0.0057	0.0070
$\Delta S/(Nk)$	0.17	0.887	0.508	0.62
η_2	0.36	0.542	0.455	0.499
τ	4.0	3.9	3.9	3.9
$(dT_{NI}/dP)_{P=1 \text{ atm}}/(\text{K/kbar}\|)$	48	175	30	31

† For the references from which these values were taken, see Savithramma & Madhusudana (1982b).
‡ Calculated by using a two-term expansion of $\langle\langle|\sin\gamma|\rangle\rangle$.
§ Calculated by using a six-term expansion of $\langle\langle|\sin\gamma|\rangle\rangle$.
‖ 1 kbar $= 10^5$ Pa.

Significantly better agreement with experimental data for PAA can be obtained by using length:width ratios less than 3, as was shown by Savithramma & Madhusudana (1980), who considered various model systems with $x = 1.0$ to $x = 2.45$. They again assumed that $v_0 = 0.230$ nm³ but chose values of v_0 and v_2 at each value of x to reproduce the experimental values of T and $v_0 \rho_{\text{nem}}$ at the N–I transition. Best results were obtained for $x = 1.75$; these values are given in column 2 of table 1. As can be seen, the overall agreement between theory and experiment is rather good.

Finally, Savithramma & Madhusudana (1982b) recently redid their calculations with the truncated two-term expansion for $\langle\langle|\sin\gamma|\rangle\rangle$ replaced by the first six terms in the exact expansion of this quantity. Their results for a system with $x = 1.75$ are given in column 4 of table 1. As can be seen, including the higher order terms in the expansion of $\langle\langle|\sin\gamma|\rangle\rangle$ increases $\Delta\rho/\rho_{\text{nem}}$, η_2, and $\Delta S/Nk$ by 23, 9.7 and 22 % respectively, but agreement with experiment is still rather good. In addition to determining the properties of model systems with various values of x at $P = 1$ atm ($ca.$ 10^5 Pa), they also calculated the properties of a system with $x = 1.75$ for pressures up to 650 MPa and obtained semiquantitative agreement with experimental data on PAA at high pressures.

(b) Results from the generalized van der Waals theory

As noted previously, the expression for $\bar{\psi}(\Omega, \rho)$ in the generalized van der Waals theory is (4). Evaluation of the integral $I(\Omega, \Omega')$ in this equation is a rather formidable problem in solid geometry because it requires the characterization of the surface $S(\gamma)$ that separates the allowed

[63]

and forbidden values of the vector r between the centres of two molecules with orientations Ω and Ω' respectively (i.e. $S(\gamma)$ is the surface traced out by the centre of a hard spherocylinder with fixed orientation Ω' as it moves about another hard spherocylinder with fixed orientation Ω in such a way that the two particles are always in contact). $S(\gamma)$ was characterized analytically by Gelbart & Gelbart (1977), who then evaluated $I(\gamma)$ numerically for a range of values of the dimensions a and l of the spherocylindrical hard cores.

The first numerical test of the generalized van der Waals approach was reported by Baron & Gelbart (1977), who considered model systems with hard-core length:width ratios x in the range 1.0–4.2 and with attractive potentials

$$v_{\text{att}}(r, \gamma) = -C_{\text{iso}}/r^6 - C_{\text{aniso}} \cos^2(\gamma)/r^6, \tag{15}$$

where C_{iso} and C_{aniso} are positive constants. To avoid solving the nonlinear integral equation (10), they introduced two new approximations, namely

$$f(\Omega) = \cosh(\alpha \cos\theta) \Big/ \int_0^\pi \cosh(\alpha \cos\theta) \sin\theta \, d\theta, \tag{16}$$

where α is determined by minimization of the free energy, and

$$\langle\langle I(\gamma) \rangle\rangle = A_0 + A_2 \eta_2^2. \tag{17}$$

Equation (16) is a one-parameter variational approximation for $f(\Omega)$ first suggested by Onsager (1949); (17) is obtained by expanding $I(\gamma)$ in even-order Legendre polynomials, averaging term by term, and then truncating the resulting series after its second term. Gelbart & Gelbart (1977) showed that the latter is a good approximation if (16) is valid and η_2 is of order 0.5. However, the very large values of η_2 actually obtained by Baron & Gelbart cast considerable doubt on the validity of this truncation. More importantly, the use of the Onsager approximation (16) clearly introduced substantial error into the calculations, as can be seen by comparing Baron & Gelbart's results for hard spherocylinders with $x = 3$ at the N–I transition ($\eta_2 = 0.96$, $v_0 \rho_{\text{nem}} = 0.707$) with the corresponding numbers obtained by Cotter (1979a) with the use of the numerical solution of the scaled particle integral equation for $f(\Omega)$ ($\eta_2 \approx 0.64$, $v_0 \rho_{\text{nem}} \approx 0.55$). Thus although Baron & Gelbart reported some interesting trends in T_{NI}, ΔS_{NI} and the densities of the coexisting phases at T_{NI} as v_0, x and C_{iso} are varied, it is clear that their results do not constitute a definitive test of the generalized van der Waals theory. Further calculations with a more accurate representation of $f(\Omega)$ were clearly needed.

To provide a more accurate test of the generalized van der Waals approach, I have quite recently carried out a series of calculations based on model systems with hard-core length:width ratios, x, of 1.5, 2.0 and 3.0, and intermolecular attractive potentials with the simple form

$$v_{\text{att}} = -\epsilon_0/r^6 - \epsilon_2 P_2(\cos\gamma)/r^6 = -\epsilon_0\{1 + \delta P_2(\cos\gamma)\}/r^6, \tag{18}$$

where ϵ_0 and ϵ_2 are positive constants and $\delta = \epsilon_2/\epsilon_0$. The results are reported here for the first time.

With this choice for v_{att}, the integral equation (10) can be written

$$\ln f(\Omega) = \ln C - \int d\Omega' f(\Omega') \{\lambda(a, l, \rho) |\sin\gamma(\Omega, \Omega')| + (\rho/kT) I(\Omega, \Omega')\}, \tag{19}$$

where
$$I(\Omega, \Omega') = I(\gamma) = -\epsilon_0\{1 + \delta P_2(\cos\gamma)\} \int e^{-\beta v^*(r, \gamma)} r^{-6} \, dr$$

$$\equiv -\epsilon_0\{1 + \delta P_2(\cos\gamma)\} I_0(\gamma). \tag{20}$$

[64]

This nonlinear integral equation was solved numerically by expanding $\ln f(\Omega)$ in even-order Legendre polynomials through $P_{20}(\cos\theta)$, i.e. by assuming that

$$\ln f(\Omega) = \sum_{n=0}^{10} F_n P_{2n}(\cos\theta),\tag{21}$$

and solving iteratively for the coefficients F_0 to F_{10}. This requires knowing the first 11 coefficients in the Legendre expansions of $|\sin\gamma|$ and $I_0(\gamma)$, i.e. the first 11 coefficients B_n and D_n, where

$$|\sin\gamma| = \sum_{n=0}^{\infty} B_n P_{2n}(\cos\gamma)\tag{22}$$

and

$$I_0(\gamma) = \sum_{n=0}^{\infty} D_n P_{2n}(\cos\gamma).\tag{23}$$

The B_n are easily determined analytically; the D_n were obtained by numerical integration with the use of the two-dimensional integral representation of $I_0(\gamma)$ derived by Gelbart & Gelbart (1977). To calculate the reduced density $v_0\rho$ and the coefficients F_n at a particular temperature T and external pressure P_{ext}, one can solve (19) simultaneously with

$$\Pi(T, v_0\rho) = P_{\text{ext}}v_0/kT,\tag{24}$$

where $\Pi(T, v_0\rho)$ is the van der Waals expression for Pv_0/kT, namely

$$\Pi = \frac{v_0\rho[1 + v_0\rho + \frac{2}{3}(v_0\rho)^2(1 + q - \frac{1}{2}q^2) + 2rv_0\rho\{1 + \frac{1}{3}(1 + 2q)v_0\rho\}\langle\langle|\sin\gamma|\rangle\rangle]}{(1 - v_0\rho)^3}$$
$$- \frac{\epsilon_0(v_0\rho)^2}{2v_0kT}\langle\langle I(\gamma)\rangle\rangle.\tag{25}$$

Convergence of the iterative procedure is, however, rather slow. Much faster convergence can be obtained by fixing P_{ext} and the coefficient F_1 (rather than P_{ext} and T) and solving for T, $v_0\rho$ and the remaining F_n by using (19), (24) and the exact relation

$$F_1 = -\{\lambda(a, l, \rho)B_1 + \rho D_1/kT\}\eta_2.\tag{26}$$

(This can be derived by substituting the Legendre expansions for $\ln f(\Omega)$, $|\sin\gamma|$, and $I(\gamma)$ in (19), integrating over Ω' term by term, and then equating the coefficients of $P_2(\cos\theta)$ on each side of the resulting equation.) The latter was the procedure actually used in the computations. Once the values of F_2 to F_{10}, η_2 to η_{20}, $v_0\rho$ and T were determined for a nematic phase with the specified P and F_1, the chemical potential μ, entropy S, Helmholtz free energy A, and internal energy U were calculated. Equation (25) with $\langle\langle|\sin\gamma|\rangle\rangle = B_0 = \frac{1}{4}\pi$ and $\langle\langle I(\gamma)\rangle\rangle = D_0$ was then solved iteratively for the density of an isotropic phase with the same T and P as the nematic phase in question and μ, S, A and U were calculated for this isotropic phase. The N–I phase transition was located by searching for the value of F_1 such that μ_{nem} and the corresponding μ_{iso} are equal. Finally, the quantity $\tau = -\rho(\partial\eta_2/\partial\rho)_T/T(\partial\eta_2/\partial T)_\rho$ was evaluated by numerical differentiation and $(dT_{\text{NI}}/dP)_{P=1\text{atm}}$ was evaluated by using the Clausius–Clapeyron equation. The results obtained in this manner are summarized in tables 2–4.

The properties of three model systems with $x = 1.5$, 2 and 3 respectively at the N–I transition at $P = 1$ atm are given in table 2. In each case, v_0 was taken to be 0.230 nm^3, ϵ_2 was set equal to zero (i.e., v_{att} was taken to be isotropic) and ϵ_0 was chosen to obtain an N–I transition temperature of approximately 409 K. The relevant experimental data for PAA are again given for purposes of comparison. As can be seen, the agreement with experiment is atrociously poor when $x = 3$, still rather poor when $x = 2$, and respectable, but not terribly impressive, when $x = 1.5$. Clearly, in

TABLE 2. VALUES OF VARIOUS QUANTITIES AT THE N–I PHASE TRANSITION AS PREDICTED BY THE GENERALIZED VAN DER WAALS THEORY WITH $v_{att} = -\epsilon_0/r^6$

($P = 1$ atm in all cases.)

quantity	experimental data for PAA	theoretical predictions		
		$x = 3$	$x = 2$	$x = 1.5$
$\{v_0/(kv_0a^3)\}/10^4$ K	—	7.60	7.73	10.7
T_{NI}/K	409	408.6	408.4	408.8
$v_0 \rho_{nem}$	0.62	0.450	0.456	0.559
$\Delta\rho/\rho_{nem}$	0.0035	0.481	0.116	0.015
$\Delta S/(Nk)$	0.17	6.99	2.65	1.07
η_2	0.36	0.974	0.816	0.607
η_4	0.07	0.919	0.572	0.284
τ	3.9	1.61	1.65	1.64
$(dT_{NI}/dP)_{P=1\,atm}/(K/kbar)$	48	497	184	44

its present form the generalized van der Waals theory greatly exaggerates the degree of orientational order in the nematic phase and the strength of the first-order N–I transition, even when an *isotropic* attractive potential v_{att} is used. The mathematical origins of this behaviour can be inferred from table 3, which lists the values of the coefficients F_n in the expansion of $\ln f(\Omega)$, the order parameters η_{2n}, and the coefficients D_n in the expansion of $I(\gamma)$ for these three model systems. When $\delta = 0$, the mean field pseudo-potential is given by

$$\overline{\psi}(\Omega,\rho) = -\rho\epsilon_0 \int f(\Omega') I_0(\gamma) \, d\Omega'$$

$$\approx -\rho\epsilon_0 \sum_{n=0}^{10} D_n \eta_{2n} P_{2n}(\cos\theta). \qquad (27)$$

It is therefore clear that the pseudo-potentials calculated from the generalized van der Waals theory are much more strongly orientation dependent than the sort of empirical $\overline{\psi}$'s that yield the best agreement with experiment. This very strong favouring of alignment results in large values of the order parameters η_2, η_4, etc., in the nematic phase and these high order parameters, together with the relatively large values of D_2, D_4, etc., yield large order-dependent terms in the internal energy U of the system. To be slightly more quantitative, let us define an anisotropy parameter by the relation

$$\xi = \{U(1) - U(0)\}/U(0), \qquad (28)$$

where $U(1)$ and $U(0)$ are, respectively, the internal energies of a perfectly ordered nematic phase and of an isotropic phase at density ρ and temperature T. For the three model systems described in tables 2 and 3, with $x = 3$, 2 and 1.5, the values of ξ are 0.679, 0.365 and 0.133 respectively. On the other hand, for the model systems described in table 1 (with empirical pseudo-potentials), $\xi = 0.080$ ($x = 3$) and $\xi = 0.032$ ($x = 1.75$).

The effects of adding an anisotropic term proportional to $P_2(\cos\gamma)$ to v_{att} can be seen in table 4, which compares model systems with $x = 1.5$ and four values of δ: 0, 0.01, 0.05 and 0.10. In each case, ϵ_0 was chosen to yield $T_{NI} \approx 409$ K. From this table it is clear that, on the whole, the agreement between theory and experiment becomes worse as δ increases. This is not surprising because the van der Waals mean field potential is already too strongly orientation-dependent when $\delta = 0$.

TABLE 3. THE COEFFICIENTS F_n AND D_n AND THE ORDER PARAMETERS η_{2n} FOR THE MODEL SYSTEMS IN TABLE 2

(F_n is the coefficient of $P_{2n}(\cos\theta)$ in the Legendre expansion of $\ln f(\Omega)$ and D_n is the coefficient of $P_{2n}(\cos\gamma)$ in the Legendre expansion of $I(\gamma)$.)

	$x = 3$			$x = 2$			$x = 1.5$		
n	F_n	$a^3 D_n$	η_{2n}	F_n	$a^3 D_n$	η_{2n}	F_n	$a^3 D_n$	η_{2n}
0	—	0.16984	1.00000	—	0.22688	1.00000	—	0.30499	1.00000
1	5.8219	0.04191	0.97429	3.83640	0.04234	0.81598	2.69633	0.02556	0.60653
2	2.47460	0.02554	0.91931	0.95666	0.01666	0.57174	0.32925	0.00682	0.28467
3	1.32693	0.01582	0.84305	0.29929	0.00836	0.36157	0.05784	0.00299	0.11364
4	0.76661	0.01044	0.75404	0.10373	0.00494	0.21295	0.01166	0.00167	0.04081
5	0.46425	0.00730	0.65981	0.03811	0.00325	0.11911	0.00249	0.00107	0.01363
6	0.29082	0.00536	0.56627	0.01444	0.00229	0.06410	0.00055	0.00074	0.00432
7	0.18648	0.00409	0.47764	0.00560	0.00170	0.03348	0.00012	0.00055	0.00132
8	0.12125	0.00321	0.39660	0.00220	0.00131	0.01707	0.00003	0.00042	0.00039
9	0.07991	0.00259	0.32458	0.00088	0.00105	0.00854	6×10^{-6}	0.00033	0.00011
10	0.05300	0.00213	0.26210	0.00036	0.00087	0.00042	1×10^{-6}	0.00027	0.00003

TABLE 4. VALUES OF VARIOUS QUANTITIES AT THE N–I PHASE TRANSITION AS PREDICTED BY THE GENERALIZED VAN DER WAALS THEORY WITH $v_{\mathrm{att}} = -(\epsilon_0/r^6)\{1 + \delta P_2(\cos\gamma)\}$

($P = 1$ atm in all cases.)

quantity	experimental data for PAA	theoretical predictions for $x = 1.5$			
		$\delta = 0$	$\delta = 0.01$	$\mu = 0.05$	$\delta = 0.10$
$\{\epsilon_0/(kv_0 a^3)\}/10^4$ K	—	10.7	9.92	7.83	6.37
$T_{\mathrm{NI}}/$K	409	408.8	408.7	408.8	408.8
$v_0 \rho_{\mathrm{nem}}$	0.62	0.559	0.547	0.506	0.468
$\Delta\rho/\rho_{\mathrm{nem}}$	0.0035	0.0153	0.0165	0.0230	0.0347
$\Delta S/(Nk)$	0.17	1.07	1.04	1.01	1.07
η_2	0.36	0.607	0.593	0.569	0.567
η_4	0.07	0.284	0.269	0.239	0.233
τ	3.9	1.64	1.57	1.39	1.28
$(\mathrm{d}T_{\mathrm{NI}}/\mathrm{d}P)_{P=1\,\mathrm{atm}}/$(K/kbar)	48	44	50	78	121

4. CONCLUSIONS AND FUTURE PROSPECTS

(a) Discussion of results

As is clear from the results presented in §3, the van der Waals approach can yield rather good agreement with experiment if one uses an empirical mean field potential with two adjustable parameters v_0 and v_2 and one chooses a hard-core length:width ratio somewhat smaller than one might guess a priori. On the other hand, when one uses the generalized van der Waals theory, in which the mean field pseudo-potential $\bar\psi(\Omega,\rho)$ is calculated from an assumed intermolecular pair potential in a self-consistent manner, the agreement between theory and experiment is much less impressive. All of the idealizations and approximations inherent in the generalized van der Waals theory no doubt contribute to its quantitative deficiencies. In my opinion, however, in order of decreasing probable importance, the most serious defects of the theory in its present form are: (1) the neglect of short-range orientational order in the evaluation of $\bar\psi(\Omega,\rho)$, (2) the use of cylindrically symmetric molecular hard cores, and (3) the use of scaled particle theory to evaluate $A^*\{f(\Omega)\}$. The evidence suggesting that the use of scaled particle theory may be a significant source of error was presented in §2. The reasons for believing that the neglect of short-range order

[67]

and the use of spherocylindrical hard cores have even more serious consequences are discussed in turn below.

In §2 it was noted that all short-range translational and orientational order is neglected in the evaluation of $\bar{\psi}(\Omega, \rho)$, except for the enforcement of the requirement that molecular hard cores not overlap. More precisely, in performing the averaging over r and Ω' in (4), it is assumed that the density of molecules with orientation Ω' at position r relative to the centre of a molecule with orientation Ω has only two values: it is equal to zero within the surface $S(\gamma)$ on which $e^{-\beta v^*(r,\gamma)}$ changes from 0 to 1; otherwise it is equal to $\rho f(\Omega')$, even for values of r just outside $S(\gamma)$. Given the success of the van der Waals theory of simple liquids, it would appear that the neglect of short-range translational order in the calculation of the mean field potential is a reasonable approximation. On the other hand, neglecting the strong short-range orientational order that no doubt exists in both nematic and isotropic phases of rod-like molecules may well have serious thermodynamic consequences. Intuitively one would expect the neglect of short-range orientational correlations when evaluating $\bar{\psi}$ to lead to (1) an overestimate of the degree of long-range orientational order in the nematic phase (in the absence of short-range orientational order, the only way that neighbouring molecules can be strongly aligned and thus minimize their energy of interaction is for there to be very strong long-range orientational order present), (2) an energetic 'discrimination' against the isotropic phase, and (3) an exaggeration of the difference between a nematic and an isotropic phase at a given T and ρ. To obtain a rough estimate of the magnitude of these effects I have redone the van der Waals calculations for two model systems with $x = 2$ and $x = 3$ respectively, using a very crude and highly arbitrary procedure for incorporating the effects of short-range orientational order into the evaluation of $\bar{\psi}(\Omega, \rho)$. Specifically, the procedure used when averaging $v_{att}(r, \Omega, \Omega')$ over r and Ω' was to define a spherocylindrical volume of radius $2a(1 + \alpha)$ and cylindrical length l surrounding the molecule with orientation Ω and then to assume that all molecules whose centres lie within this volume also have orientation Ω. (Outside this spherocylindrical volume, the density of molecules with orientation Ω' was again assumed to be $e^{-\beta v^*} \rho f(\Omega')$.) For both systems considered, α was set equal to $l/(8a)$, which means that the radius $2a(1 + \alpha)$ is midway between the respective distances of closest approach of two parallel and two perpendicular hard spherocylinders of radius a and cylindrical length l. The results of these revised calculations are given in table 5, from which it is clear that the incorporation of short-range order – however crudely and arbitrarily it was done – leads to dramatically improved agreement with experiment. This argues quite strongly that the neglect of short-range orientational order in the evaluation of $\bar{\psi}(\Omega, \rho)$ is a serious source of error in the generalized van der Waals theory.

Real nematogenic molecules clearly do not have cylindrically symmetric shapes. None the less, given the statistically free rotation about the molecular long axis in nematic liquids, it might seem that the use of spherocylindrical molecular hard cores is a reasonable approximation. However, calculations by a number of authors (Shih & Alben 1972; Straley 1974; Luckhurst *et al.* 1975), with the use of a variety of approximate theoretical approaches, suggest that the assumption of cylindrical molecular symmetry leads to significant overestimates of the degree of order in the nematic phase and of the discontinuities in ρ, H, S, η_2, etc., at the N–I transition. Most recently, for example, Gelbart & Barboy considered systems of hard ellipsoids of revolution (Gelbart & Barboy 1979) and of hard parallelepipeds (Gelbart & Barboy 1980) with principal axes of length a, b and c, constrained to lie with their longest axis parallel to one of the six axial directions of a space-fixed Cartesian coordinate system. In both instances they calculated the nematic order parameter $\eta_{2,\mathrm{NI}}$ and the relative density change $\Delta\rho/\rho_{\mathrm{nem}}$ at the N–I transition as a function of b/a with c/a held equal to 5. Both $\eta_{2,\mathrm{NI}}$ and $\Delta\rho/\rho_{\mathrm{nem}}$ were found to decrease with increasing b/a, becoming zero at $b/a = b^*/a \approx 2.3$. For $b > b^*$, the ordered phase had $\eta_2 < 0$; i.e. it was a uniaxial phase in which the shortest molecular axis was oriented parallel to the uniaxial symmetry axis. Based on these results, the authors argued that the very small values of $\Delta\rho$, ΔS, etc., observed at the N–I transition in common nematogens result from a nearly equal balance between the tendencies to form a hard-rod-like and a hard-plate-like uniaxial phase. If this is correct, then the use of cylindrically symmetric hard cores in the van der Waals theory probably makes a major contribution to the quantitative deficiences observed.

TABLE 5. VALUES OF VARIOUS QUANTITIES AT THE N–I PHASE TRANSITION AS PREDICTED BY THE GENERALIZED VAN DER WAALS THEORY, WITH AND WITHOUT 'CORRECTION' FOR THE EFFECTS OF SHORT-RANGE ORIENTATIONAL ORDER

(In all cases, $v_{\mathrm{att}} = -\epsilon_0/r^6$ and $P = 1$ atm.)

quantity	experimental data for PAA	theoretical predictions for $x = 3$		theoretical predictions for $x = 2$	
		uncorrected	'corrected' $\alpha = 0.50$	uncorrected	'corrected' $\alpha = 0.25$
$\{\epsilon_0/(kv_0a^3)\}/10^4$ K	—	7.60	10.0	7.73	12.0
T_{NI}/K	409	408.6	408.7	408.4	408.8
$v_0\rho_{\mathrm{nem}}$	0.62	0.450	0.493	0.456	0.556
$\Delta\rho/\rho_{\mathrm{nem}}$	0.0035	0.481	0.040	0.116	0.012
$\Delta S/(Nk)$	0.17	6.99	1.01	2.65	0.599
η_2	0.36	0.974	0.608	0.816	0.476
η_4	0.07	0.919	0.265	0.572	0.149
τ	4.0	1.61	7.46	1.65	3.53
$(\mathrm{d}T_{\mathrm{NI}}/\mathrm{d}P)_{P=1\,\mathrm{atm}}/(\mathrm{K/kbar})$	48	497	141	184	61

(b) Suggested improvements

Based on the discussion in the previous subsection, it appears that the most promising ways to attempt to improve the generalized van der Waals theory would be (i) to incorporate short-range orientational order into the theory in some approximate but self-consistent manner, (ii) to use molecular hard cores with lower than cylindrical symmetry, and (iii) to devise a method for evaluating the free energy functional $A^*\{f(\Omega)\}$ more accurately. Some brief comments concerning each of these suggested improvements are given in the following paragraphs.

(i) Incorporation of short-range orientational order

Devising a way to 'build in' short-range orientational correlations in a tractable *and* self-consistent manner is clearly a very difficult but worthwhile task. One possible way to proceed would be to replace (4) for the pseudo-potential $\overline{\psi}(\Omega,\rho)$ by

$$\overline{\psi}(\Omega,\rho) = \rho \int f(\Omega')\,\mathrm{d}\Omega' \int \mathrm{d}\boldsymbol{r}\, \mathrm{e}^{-\beta v^*(r,\gamma)} y(\boldsymbol{r},\gamma)\, v_{\mathrm{att}}(\boldsymbol{r},\Omega,\Omega'), \qquad (29)$$

where $\mathrm{e}^{-\beta v^*}y$ represents some (necessarily crude) approximation to the hard-core pair correlation function $g^{(2)}(\rho,\boldsymbol{r},\Omega,\Omega')$. For the complete statistical mechanical self-consistency of the van der Waals theory to be maintained, the $y(\boldsymbol{r},\gamma)$ used could not explicitly be ρ-dependent, nor could it depend on any parameter that changes with density when $f(\Omega)$ is held constant. A reasonable functional form to use for $y(\boldsymbol{r},\gamma)$ might possibly be inferred from Monte Carlo simulations of

isotropic hard-rod fluids. If no approach of this sort proves practicable, one can always fall back
on some *ad hoc* method for mimicking the effects of short-range order – perhaps a somewhat more
sophisticated version of the scheme described in §*a*.

(ii) *Use of non-cylindrically symmetric hard cores*

The main complication that arises if one uses molecular hard cores with lower than cylindrical
symmetry in the van der Waals theory is that the evaluation of the integral

$$I(\Omega, \Omega') = \int d\boldsymbol{r}\, e^{-\beta v^*} v_{\mathrm{att}}$$

becomes much more difficult. In particular, the characterization of the surface $S(\Omega, \Omega')$ on
which $e^{-\beta v^*}$ changes from 0 to 1 becomes a *quite* difficult problem in solid geometry. Whether
or not $I(\Omega, \Omega')$ can be evaluated in practice depends on how clever one is in choosing a hard-core
shape. In my opinion, a hard ellipsoid would not be a good choice for the hard-core shape; a more
promising candidate would be what can be called a capped parallelepiped, i.e. a parallelepiped
with edges of lengths a, b, and l capped on the two faces of area al by a half spherocylinder of
radius a and cylindrical length l and capped on the two faces of area ab by a half spherocylinder
of radius a and cylindrical length b. (In this limit, $b = 0$, this becomes a spherocylinder of radius a
and cylindrical length l.) The advantage of this shape is that in moving one such object about
another to determine $S(\Omega, \Omega')$, one is always sliding a point on a spherocylinder or a plane over
a planar or spherocylindrical surface. This leads to less complicated geometry than does sliding
one ellipsoid over another. (As a trivial example of the differences encountered, note that the
mutual exclusion volume of two aligned spherocylinders is a spherocylinder whereas the mutual
exclusion volume of two ellipsoids is not an ellipsoid.) Based on some preliminary estimates, it
seems that it *may* be possible to apply the van der Waals approach to a model system with hard
cores in the shape of such a capped parallelepiped or 'stretched spherocylinder'.

(iii) *Other methods of evaluating $A^*\{f(\Omega)\}$*

In principle one could derive a more accurate hard-core free energy functional $A^*\{f(\Omega)\}$ by
solving one of the common integral or integro-differential equations for the pair correlation
function of the hard-core system and then evaluating $A^*\{f(\Omega)\}$ by standard statistical mechanical
manipulations. In practice, however, the probability of doing this successfully for a system of
hard rods seems very small at present. One must therefore look for some approximate, tractable
method of evaluating $A^*\{f(\Omega)\}$ that is more accurate than scaled particle theory. Two possibilities have so far been suggested.

Savithramma & Madhusudana (1980) evaluated $A^*\{f(\Omega)\}$ for hard spherocylinders by using
an extension of a method developed by Andrews (1975) for calculating the hard-sphere equation
of state. In this approach the reciprocal of the thermodynamic activity, a^{-1}, is identified with the
probability p that an additional hard particle can be added successfully to the system at some
arbitrary point; p is then evaluated by using simple probabilistic arguments, in which the total
free volume V_f in the system is written $V_f = V - N\omega(\rho)$, where $\omega(\rho)$ is determined by recourse to
computer 'experiments'. Savithramma & Madhusudana expressed $\omega(\rho)$ as a seven-term power
series in ρ whose coefficients were evaluated by demanding that $\omega = 2\sqrt{3}\, v_0$ at the close-packed
density and that the coefficient of ρ^n ($n = 3$ to 8) in the density expansion of Pv_0/kT for an isotropic
state be equal to the virial coefficient B_n determined by Monte Carlo calculation. Using

$A*\{f(\Omega)\}$ calculated in this manner, the empirical pseudo-potential (2), and the two-term expansion of $\langle\langle|\sin\gamma|\rangle\rangle$, they determined the properties of various model systems with $x = 1.0$ to 2.90 at the N–I transition, obtaining somewhat better agreement with experiment than that produced by scaled particle theory. Given the highly approximate free volume arguments used in deriving the activity a, this improvement clearly results from 'building in' the virial equation of state for isotropic hard spherocylinders. The method is thus limited to systems for which a large number of virial coefficients are available for the isotropic hard-core system. In such cases, the 'Andrews' $A*\{f(\Omega)\}$ could readily be used in place of the scaled particle free energy functional in the generalized van der Waals theory. I doubt, however, that this would by itself yield greatly improved results.

Another alternative to scaled particle theory that has been proposed is the y-expansion technique developed by Barboy & Gelbart (1979, 1980). In this approach, the equation of state is expanded in a power series in the variable $y = v_0\rho/(1-v_0\rho)$, i.e.

$$Pv_0/kT = \sum_{n\geqslant 1} C_n y^n, \tag{30}$$

where the coefficients C_n are related to the usual virial coefficients B_2, B_3, B_4, etc., by

$$B_{n+1} = \sum_{k=1}^{n+1} \binom{n}{k-1} C_k. \tag{31}$$

For isotropic fluids of hard particles with a variety of shapes, the y-expansion – unlike the ordinary virial expansion – has been shown to converge rapidly enough to calculate the equation of state accurately at liquid-like densities. It has been used to study the N–I phase transition in systems of hard parallelepipeds restricted to six discrete orientations (parallel to the $\pm x$, $\pm y$, $\pm z$ axes of a space-fixed Cartesian coordinate system) and to calculate the equation of state of an isotropic fluid of hard spherocylinders. (In the latter instance, truncating the expansion after the y^3 term yielded results comparable with those of scaled particle theory.) The difficulty in applying this method to nematic phases of rod-like particles allowed to adopt all orientations is that the virial coefficients B_2, B_3, B_4, etc., are generally not available and very hard to calculate. For a system of hard rods with orientational distribution function $f(\Omega)$,

$$B_n = \int \cdots \int f(\Omega_1) f(\Omega_2) \ldots f(\Omega_n) \beta_{n-1}(\Omega_1, \ldots, \Omega_n)\, \mathrm{d}\Omega_1 \ldots \mathrm{d}\Omega_n, \tag{32}$$

where β_{n-1} is the $(n-1)$th irreducible cluster integral for n hard rods with fixed orientations $\Omega_1, \Omega_2, \ldots, \Omega_n$, respectively. $\beta_1(\Omega_1, \Omega_2)$ has been derived only for hard spherocylinders, cylinders, prolate ellipsoids and oblate ellipsoids (Onsager 1949; Isihara 1951); $\beta_2(\Omega_1, \Omega_2, \Omega_3)$ has not been derived for hard rods of any shape. Moreover, it seems highly unlikely that analytical expressions for β_2, β_3, etc., will ever be derived. As noted by Gelbart & Barboy, however, it *may* be possible to derive accurate approximate expressions for $\beta_2(\Omega_1, \Omega_2, \Omega_3)$ at least. If this proves to be so and if the y-expansion converges sufficiently rapidly in the nematic phase, this approach may turn out to be a more accurate way to evaluate $A*\{f(\Omega)\}$. At the moment this is still an open question.

(c) Extensions of the theory

Cotter & Wacker (1978a) extended the van der Waals theory to nematogenic solutions, i.e. solutions that exhibit a stable nematic mesophase over some range of temperature and composition. The extended theory is applicable to mixtures of any number of components with spherocylindrical–spherical molecular hard cores. So far, however, it has been applied only to binary mixtures with effectively spherical solute molecules and rod-like solvent molecules (Cotter & Wacker 1978b), by using an empirical mean field pseudo-potential for the solvent molecules. (Solute and solvent molecular dimensions were estimated from tabulated van der Waals volumes, and all other energy parameters were calculated from ratios of heats of vaporization.) The temperature–mole-fraction phase diagrams calculated from the theory are in rather good agreement with experimental data (Martire et al. 1976) for the systems CCl_4-DHAB, $(CH_3)_4$Sn-MBBA, $(C_2H_5)_4$Sn-MBBA, $(n\text{-}C_3H_7)_4$Sn-MBBA, and $(n\text{-}C_4H_9)_4$Sn-MBBA.

More recently, Gelbart & Ben-Shaul (1982) have extended the theory to nematic mesophases subject to elastic deformations. For each of the principal elastic constants (i.e. splay, bend and twist), they derived an expansion of the form

$$k = C_{22}\eta_2^2 + C_{24}\eta_2\eta_4 + C_{44}\eta_4^2 + \cdots,$$ (33)

which they truncated after the term proportional to $\eta_2\eta_4$ (η_n is equal to $P_n(\cos\theta)$, as usual). These expansions were then used to investigate the dependence of the elastic constants on various molecular parameters. When experimental values are used to determine $\eta_2(T)$ and $\eta_4(T)$, the predictions of the theory are in reasonably good agreement with experimental observations.

Also quite recently, Savithramma & Madhusudana (1982a) have extended the van der Waals approach to systems of disc-like nematogens, using molecular hard cores in the shape of a right circular cylinder of radius a and length l and the empirical pseudo-potential (2). This model system was shown to exhibit a hard-rod-like nematic mesophase ($\eta_2 > 0$) when $x = l/2a > 1$ and a hard-disc-like nematic mesophase ($\eta_2 < 0$) when $x < 1$. Moreover, when various properties of the system at the N–I transition (e.g. $\Delta\rho/\rho_{nem}$ and $\Delta U/(NkT)$) were plotted against x, the curves were approximately symmetric about $x = 1$, suggesting that the properties of the N–I transition for disc-like nematogens are comparable in many respects with those for rod-like nematogens.

Future extensions of the van der Waals theory that would in my opinion be particularly desirable include (1) extending the theory to allow for the possibility of smectic as well as nematic ordering, and (2) modifying the theory to take into account the flexibility of the molecular 'end chains' of real nematogens or smectogens. As a preliminary endeavour before attempting either of these extensions, a student of mine, L. Petrone, is working at present on a lattice version of the van der Waals theory that incorporates end-chain flexibility in a fairly realistic manner and can treat smectic ordering.

(d) Concluding remarks

In conclusion, I should like to return to a point emphasized in the introduction, namely that the van der Waals theory is intended to be a 'chemist's' theory to be used to study the relation between molecular structure and mesomorphic behaviour. In my opinion there are two main requirements that a successful theory of this sort must meet: (i) although quantitative agreement between theory and experiment is certainly not required, the agreement must be good enough so that one can believe qualitative trends and explanations suggested by the theory – even rather subtle ones; (ii) it must be possible to relate the parameters of the model system to the molecular size, shape, polarity, polarizability, etc., of a particular mesogen in some reasonably straightforward manner. The van der Waals approach with the empirical pseudo-potential (2) satisfies requirement (i) nicely, but falls far short of meeting requirement (ii) because there is no clear way to relate the value of the parameter v_2 to the characteristics of a particular molecule. On the other hand, the generalized van der Waals theory in its present state has some trouble with requirement (i) owing to its severe overestimate of the orientation dependence of the mean field potential. This latter problem can, however, clearly be corrected. (The only question is whether it can be corrected in a less crude and arbitrary manner than that described in §a.) Furthermore, although the generalized van der Waals approach does not completely satisfy requirement (ii) at present, it comes closer, in my opinion, than any other molecular theory proposed to date. In short, although much remains to be done, a start has clearly been made toward the goal of developing a successful 'chemical' theory of thermotropic mesomorphism. Modifying the van der Waals theory to take into account molecular flexibility, non-cylindrically symmetric molecular shapes, and the existence of smectic order would clearly all be large steps toward this goal.

REFERENCES

Andrews, F. C. 1975 *J. chem. Phys.* **62**, 272–275.
Barboy, B. & Gelbart, W. M. 1979 *J. chem. Phys.* **71**, 3053–3062.
Barboy, B. & Gelbart, W. M. 1980 *J. statist. Phys.* **22**, 709–742.
Baron, B. A. & Gelbart, W. M. 1977 *J. chem. Phys.* **67**, 5795–5801.
Cotter, M. A. 1974 *Phys. Rev.* A **10**, 625–636.
Cotter, M. A. 1977a *J. chem. Phys.* **66**, 1098–1106.
Cotter, M. A. 1977b *J. chem. Phys.* **66**, 4710–4711.
Cotter, M. A. 1977c *J. chem. Phys.* **67**, 4268–4270.
Cotter, M. A. 1979a In *The molecular physics of liquid crystals* (ed. G. R. Luckhurst & G. W. Gray), pp. 169–180. London: Academic Press.
Cotter, M. A. 1979b In *The molecular physics of liquid crystals* (ed. G. R. Luckhurst & G. W. Gray), pp. 181–189 London: Academic Press.
Cotter, M. A. & Martire, D. E. 1970a *J. chem. Phys.* **52**, 1909–1919.
Cotter, M. A. & Martire, D. E. 1970b *J. chem. Phys.* **53**, 4500–4511.
Cotter, M. A. & Wacker, D. C. 1978a *Phys. Rev.* A **18**, 2669–2675.
Cotter, M. A. & Wacker, D. C. 1978b *Phys. Rev.* A **18**, 2676–2682.
Gelbart, W. M. & Baron, B. A. 1977 *J. chem. Phys.* **66**, 207–213.
Gelbart, W. M. & Barboy, B. 1979 *Molec. Cryst. liq. Cryst.* **55**, 209–226.
Gelbart, W. M. & Barboy, B. 1980 *Acct. chem. Res.* **13**, 290–296.
Gelbart, W. M. & Ben-Shaul, A. 1982 *J. chem. Phys.* **77**, 916–933.
Gelbart, W. M. & Gelbart, A. 1977 *Molec. Phys.* **33**, 1387–1398.
Isihara, A. J. 1951 *J. chem. Phys.* **19**, 1142–1147.
Lasher, G. 1970 *J. chem. Phys.* **53**, 4141–4146.
Longuet-Higgins, H. C. & Widom, B. 1964 *Molec. Phys.* **8**, 549–556.
Luckhurst, G. R. 1979 In *The molecular physics of liquid crystals* (ed. G. R. Luckhurst and G. W. Gray), pp. 85–119. London: Academic Press.
Luckhurst, G. R., Zannoni, C., Nordiv, P. L. & Segre, U. 1975 *Molec. Phys.* **30**, 1345–1351.
Maier, W. & Saupe, A. 1959 *Z. Naturf.* **14**a, 882–889.
Maier, W. & Saupe, A. 1960 *Z. Naturf.* **15**a, 287–292.
Martire, D. E., Oweimreen, G. A., Ågren, G. I., Ryan, S. G. & Peterson, H. T. 1976 *J. chem. Phys.* **64**, 1456–1464.
McColl, J. R. & Shih, C. S. 1972 *Phys. Rev. Lett.* **29**, 85–87.
Monson, P. A. & Rigby, M. 1978 *Molec. Phys.* **35**, 1337–1342.
Onsager, L. 1949 *Ann. N.Y. Acad. Sci.* **51**, 627–643.
Reiss, H., Frisch, H. L. & Lebowitz, J. L. 1959 *J. chem. Phys.* **31**, 369–380.
Savithramma, K. L. & Madhusudana, N. V. 1980 *Molec. Cryst. liq. Cryst.* **62**, 63–80.
Savithramma, K. L. & Madhusudana, N. V. 1982a *Molec. Cryst. liq. Cryst.* (In the press.)
Savithramma, K. L. & Madhusudana, N. V. 1982b *Molec. Cryst. liq. Cryst.* (In the press.)
Shih, C. S. & Alben, R. 1972 *J. chem. Phys.* **57**, 3055–3061.
Straley, J. P. 1974 *Phys. Rev.* A **10**, 1881–1887.
Timling, K. 1974 *J. chem. Phys.* **61**, 465–469.
Vieillard-Baron, J. 1974 *Molec. Phys.* **28**, 809–818.

Discussion

J. D. Litster (*M.I.T., Cambridge, Massachusetts, U.S.A.*) I have two questions.

(1) How much better is Professor Cotter's approach than the simple Maier–Saupe model?

(2) Can Professor Cotter estimate the difference between the first order transition temperature and the extrapolated second order transition from short range order measurements in the isotropic phase?

Martha A. Cotter.

(1) If one uses the van der Waals approach with an empirical mean-field potential and treats the hard-core length:width ratio and the energy parameters v_0 and v_2 as adjustable parameters, one obtains agreement with experiment comparable with that obtained from the simple Maier–Saupe theory – better agreement in some respects, worse in others. I do not think this is of great significance, however. As I have tried to indicate, what I think the van der Waals theory is better for is studying the relation between molecular structure and mesophase stability and properties. On the other hand, the van der Waals approach is no doubt worse than the simple Maier–Saupe approach for studying the N–I phase transition *per se* (i.e. in the context of the modern theory of phase transitions) because the former exaggerates the strength of the first-order N–I transition even more than the latter.

(2) One can no doubt calculate $T_c - T^*$ from the van der Waals theory, although I have not done so. Since the theory does not consider short-range orientational order in the isotropic phase and greatly overestimates the strength of the N–I transition, I suspect that it will also greatly overestimate $T_c - T^*$.

Lin Lei (*Institute of Physics, Chinese Academy of Sciences, Beijing, China*). We have recently extended the Landau–de Gennes theory to treat the pressure effects of nematics. Now,

$$G(T,P) = G_0(T,P) + \tfrac{1}{2}a\{T - T^*(P)\}S^2 - \tfrac{1}{3}B(P)S^3 + \tfrac{1}{4}C(P)S^4.$$

For PAA (B, C independent of P), our theory is able to explain all the existing pressure experiments (not too close to the N–A transition). Above $P = 472$ MPa and $258 < T < 251\,°$C, a re-entrant isotropic phase is predicted to exist (Lin Lei & Liu Jiagang, *Kexue Tongbao* **27**, 784 (1982); *Molec. Cryst. liq. Cryst.* (in the press)). It will be very interesting to check this result with molecular calculations. Since Professor Cotter's van der Waals theory is simple and tractable, relatively speaking, I wonder if it can be used to calculate the $T_{NI}(P)$ curve at high pressure (*ca.* 472 MPa)? If so, has this been done yet?

Martha A. Cotter. I have not calculated the N–I transition temperature at high pressures from the generalized van der Waals theory, although it can readily be done. However, Savithramma & Madhusudana (*Molec. Cryst. liq. Cryst.* (in the press)) have calculated $T_{NI}(P)$ for pressures up to 600 MPa, using the van der Waals approach with an empirical mean-field potential. They did not observe a re-entrant isotropic phase.

Phil. Trans. R. Soc. Lond. A **309**, 145–153 (1983)
Printed in Great Britain

Structural studies of nematic and smectic phases

By J. D. Litster

Department of Physics and Center for Materials Science and Engineering,
Massachusetts Institute of Technology, Cambridge, Massachusetts 02139, U.S.A.

The information about liquid crystal phases that can be obtained by light scattering and by high-resolution X-ray scattering is reviewed. Results for the nematic–smectic A transition suggest the de Gennes–McMillan model is correct, but adequate theoretical solutions to the model remain elusive. Recent results on the smectic A to smectic C transition are presented that show unambiguously that it exhibits classic mean-field behaviour and this is explained by a Ginzburg criterion argument. Some preliminary results of a study of a nematic–smectic A transition in a lyotropic material are given and indicate similarity to thermotropic materials.

1. Introduction

Nematic (N) phases of liquid crystals have long-range orientational order of molecules with rod-like anisotropy, but lack any positional order. The smectic A (SmA) and smectic C (SmC) phases have nematic order combined with a one-dimensional density wave along the director, or molecular orientation (SmA), or at an angle to it (SmC). The nematic–isotropic transition is rather well understood but the N–SmA transition remains one of the major unsolved problems in critical phenomena; it is also interesting as probably the simplest example of melting in three dimensions. The SmA–SmC transition is also interesting because it is expected to conform to the three-dimensional x–y model that is also thought to describe superfluid helium. This should have non-classical or non-mean-field behaviour, yet both classical and non-classical results have been reported by various investigators. In this paper I shall discuss results obtained by myself and my collaborators that elucidate these problems. I shall also mention some experiments on a lyotropic system of soap, water and ionic solute that show a neat soap to micellar transition quite analogous to a thermotropic N–SmA transition.

(a) Some theoretical background

Thanks to recent theoretical developments, largely arising from the work for which Kenneth Wilson received this year's Nobel Prize in Physics, we have an improved understanding of order–disorder-type phase changes. We believe that fundamental considerations like geometry and symmetry combine with thermally excited fluctuations to determine the nature of the phase changes. For most isotropic systems in four or more dimensions we believe that fluctuations are not important, so statistical mechanics calculations can be done by a simple mean-field approximation; we say that the upper marginality d^* is four. When the space dimension d is greater than d^*, classical behaviour occurs. If $d < d^*$ we have the region of critical phenomena; the fluctuations are importantly large and Wilson showed how to do statistical mechanics under these conditions. The physical idea is that fluctuations become correlated over long distances ξ so that behaviour is determined by the correlation length rather than the details of the interactions that produce order. In the Wilson, or renormalization group (r.g.), method the length scale of the problem is changed, short-wavelength fluctuations are integrated out, and thermo-dynamic quantities such as the free energy suitably rescaled. Frequently the problem becomes invariant under this procedure and a 'fixed point' has been reached. The result of the calculation then describes a system for which the correlation length ξ is infinite, which occurs at the phase transition point. Various approximate methods can then be used to calculate behaviour in the vicinity of the phase transition. Systems that all have the same mathematical fixed point should have identical physical behaviours and are in the same 'universality class'.

Finally, I should add that if $d < d^0$, where d^0 is the lower marginal dimensionality, the fluctuations are sufficiently important to prevent the establishment of order that interactions would otherwise favour. As an example, we believe that solids cannot exist in less than two dimensions. More pertinent to this paper, $d^0 = 3$ for SmA and SmC phases; this has been verified experimentally for the SmA phase (Als-Nielsen et al. 1980).

There is a rich variety of phase changes in liquid crystals and they may be studied to elucidate many of these ideas. The SmA–SmC transition has two degrees of freedom for the order parameter and should be in the 3D x–y model universality class. The de Gennes (1972)–McMillan (1971) model for an SmA is analogous to a changed superfluid and might also be thought to be in the 3D x–y class, but there are complications as we shall see presently.

2. SCATTERING FROM LIQUID CRYSTALS

Both X-rays and light can be scattered by thermal fluctuations in liquid crystals and thus used to study them. The smectic mass density waves scatter X-rays, whereas light is scattered primarily by molecular reorientations or the nematic director fluctuations; the two techniques provide different and complementary structural information about liquid crystal nematic and smectic phases.

To discuss the scattering quantitatively it is useful to have some formulae. We first consider the SmA–nematic transition. The SmA order parameter was defined by de Gennes (1972) by writing the density as

$$\rho(z) = \rho_0\{1 + \text{Re}\,(\psi e^{iq_0 z})\}, \tag{1}$$

where the director $\hat{n}$ lies along $\hat{z}$. Then ψ has two degrees of freedom, which may be taken to be the amplitude and phase of the density wave, whose wave vector is $q_0\hat{z}$. A Landau or mean-field expansion of the free energy density may be written as

$$\phi_s = \phi_0 + \tfrac{1}{2}a|\psi|^2 + \tfrac{1}{4}b|\psi|^4 + \frac{1}{2M_\parallel}|\partial_z\psi|^2 + \frac{1}{2M_\perp}|\{(\partial_x + iq_0 n_x) + (\partial_y + iq_0 n_y)\}\psi|^2 + \phi_N, \tag{2}$$

where ϕ_N is the nematic free energy for small fluctuations in the director, namely

$$\phi_N = \tfrac{1}{2}\{K_1(\nabla\cdot\hat{n})^2 + K_2(\hat{n}\cdot\text{curl}\,\hat{n})^2 + K_3(\hat{n}\times\text{curl}\,\hat{n})^2\}. \tag{3}$$

In (2), $\partial_z \equiv \partial/\partial_z$, etc., and the peculiar transverse gradient terms are required to make ϕ invariant under small rotations of the director and density wavevector together. The elastic constants in (3) were defined years ago by F. C. Frank (1958). As in any Landau theory, $a = a_0(T - T_c)$, whereas b is independent of temperature. In the SmA phase, bend ($\hat{n}\times\text{curl}\,\hat{n}$) involves compressing the density wavelength, and twist ($\hat{n}\cdot\text{curl}\,\hat{n}$) would involve a tilt angle between $\hat{n}$ and q_0; thus curl $\hat{n}$ is excluded from the SmA phase. Readers familiar with the Ginzburg–Landau (1950) model of superconductivity will recognize (2) as isomorphous with

ϕ_N playing the role of the diamagnetic $cH^2/8\pi$ term. The exclusion of curl $\hat{n}$ from the SmA phase is analogous to the Meissner effect and the director $\hat{n}$ is the analogue to the vector potential A. Analogous to fluctuation diamagnetism, the twist and bend elastic constants K_2 and K_3 diverge in the nematic phase above the N–SmA transition. This behaviour is ideally studied by light scattering from the director modes in the N and SmA phases.

More specifically, it is straightforward to show from (2) that order parameter fluctuations are described by

$$\langle \delta\psi(0)\,\delta\psi(r)\rangle = kT\chi \exp\{-\sqrt{(x^2+y^2)}/\xi_\perp - z/\xi_\parallel\}. \tag{4}$$

If we write $t = T/T_c - 1$ then the susceptibility $\chi = a^{-1}$ diverges as t^{-1} and the correlation lengths ξ as $t^{-\frac{1}{2}}$ in the mean-field calculation. More generally, $\chi = \chi_0 t^{-\gamma}$, $\xi_\parallel = \xi_\parallel^0 t^{-\nu_\parallel}$, and $\xi_\perp = \xi_\perp^0 t^{-\nu_\perp}$. These fluctuations can be studied by X-ray scattering. The intensity data for $T > T_c$ are well fitted by

$$I(q) = kT\chi/\{1 + \xi_\parallel^2(q_z - q_0)^2 + \xi_\perp^2(q_x^2 + q_y^2) + c\xi_\perp^4(q_x^2 + q_y^2)^2\}. \tag{5}$$

Except for the $c\xi_\perp^4$ term, which is required to fit the data, (5) is just the Fourier transform of (4).

There are two director modes: $\hat{n}_1$ in the plane of $\hat{n}_0 = \hat{z}$ and q, and $\hat{n}_2$ normal to $\hat{n}_0$ and $\hat{n}_1$. From (3) and (4) one may readily obtain, with q in the xz plane:

$$\langle \delta n_1^2(q)\rangle = kT\{B(q_z^2/q_x^2) + K_1 q_x^2 + K_3 q_z^2\}^{-1} \tag{6}$$

and

$$\langle \delta n_2^2(q)\rangle = kT\{D + K_2 q_x^2 + K_3 q_z^2\}^{-1}. \tag{7}$$

These are proportional to the intensity of scattered light; the spectrum contains useful information on the dynamical behaviour of fluctuations, but I shall not discuss that further. The elastic constants B and D give restoring forces for compression of the smectic density wave and tilt between the wavevector q_0 and the director $\hat{n}$, respectively. In a mean-field calculation both B and D are proportional to the square of the mean smectic order parameter $\psi_0 = \langle\psi\rangle$, while isotropic scaling laws (Jähnig & Brochard 1974) predict that B vanishes as $\xi_\parallel^{-1}$ and D as $\xi_\perp^{-1}$ in the SmA phase. The divergences of K_2 and K_3 are (Jähnig & Brochard 1974)

$$K_2 = kTq_0^2\xi_\perp^2/24\pi\xi_\parallel \tag{8}$$

and

$$K_3 = kTq_0^2\xi_\parallel/24\pi. \tag{9}$$

These equations are for the values of K_2, K_3 at long wavelength. As the correlation length grows to exceed the wavelength of the fluctuation, the divergence levels off; these non-hydrodynamic effects for $q\xi > 1$ have also been calculated in a simple mode coupling approximation. The prediction for the divergent part of K_3 with $q = q_z\hat{z}$ is (Jähnig & Brochard 1974)

$$\widetilde{K}_3 = (kTq_0^2/8q_z)\{(1 + 1/X^2)\arctan X - 1/X\}, \tag{10}$$

where $X = \frac{1}{2}q_z\xi_\parallel$. The correlation length $\xi_\parallel$ becomes long enough near the phase transition that $q\xi_\parallel$ values as large as 10 are readily obtained in a light-scattering experiment; through equation (10) one may measure $\xi_\parallel$ indirectly and compare with the direct measurements by X-ray scattering.

(a) The nematic–smectic A transition

The N–SmA transition has been studied by X-ray scattering in a number of materials. In figure 1, I show the results (Birgeneau *et al.* 1981) for butoxybenzylidene octylaniline (40.8). The most striking feature for all of the materials that have been studied is that $\xi_\parallel$ and $\xi_\perp$ diverge with different exponents. This is contrary to our usual understanding of critical phenomena,

148   J. D. LITSTER

since one length usually determines the behaviour. Light-scattering measurements of $\xi_\parallel$ (Von Känel & Litster 1981) in 40.8 analysed with (10) are in excellent agreement with the data of figure 1. In figure 2, I show $\xi_\parallel$ for butoxybenzylidene heptylaniline (40.7) obtained by X-ray and light scattering; they are also in excellent agreement. Thus it appears that the de Gennes–McMillan model and the treatment of coupling between director fluctuations and the smectic order parameter that results in (10) are valid. However, the anisotropy of the correlation lengths poses a problem. Most theoretical treatments (Lubensky 1982) predict isotropic critical behaviour ($\nu_\parallel = \nu_\perp$) and some indicate that the N–SmA transition must always be first-order to a degree

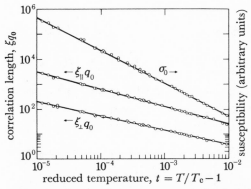

FIGURE 1. Susceptibility, σ_0, and the longitudinal and transverse correlation lengths above the SmA phase of 40.8. Data are from X-ray scattering (Birgeneau et al. 1981). The value of q_0 is 0.222 Å⁻¹.

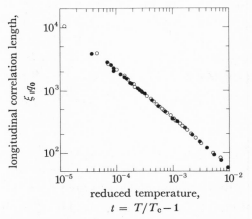

FIGURE 2. Longitudinal correlation length plotted against reduced temperature above the SmA phase of 40.7. Open circles are from light scattering, solid circles from X-ray scattering.

that is contradicted by experiment. Theories based on a dislocation mechanism for the N–SmA transition (Huberman et al. 1975; Helfrich 1978; Nelson & Toner 1981) suggest $\xi_\parallel = \xi_\perp^2$, thus $\nu_\parallel = 2\nu_\perp$. Experimental values (from Birgeneau et al. (1981) and references therein) are shown for several materials in table 1. As can be seen from the table, one does not have $\nu_\parallel = 2\nu_\perp$, either. There is an anisotropic version of the scaling hypothesis (Chen & Lubensky 1978) that predicts $\nu_\parallel + 2\nu_\perp = 2 - \alpha$, where the specific heat diverges as $|t|^{-\alpha}$. The 3D x–y model predicts $\nu_\parallel = \nu_\perp = 0.67$ and $\alpha = -0.02$. An interesting observation is that the anisotropic scaling law holds within experimental error for all materials that have been studied.

[78]

TABLE 1. CORRELATION LENGTH EXPONENTS MEASURED BY X-RAY SCATTERING
NEAR THE N–SmA TRANSITION

material†	$\nu_\perp$	$\nu_\parallel$	T_{NA}/T_{NI}
8CB	0.51	0.67	0.977
8OCB	0.58	0.71	0.963
40.8	0.58	0.70	0.958
CBOOA	0.61	0.72	0.940
$\bar{8}$S5	0.68	0.83	0.935
40.7	0.66	0.79	0.926

† Materials not previously defined are: 8CB, octyl cyanobiphenyl; 8OCB, octyloxy cyano biphenyl; CBOOA, cyanobenzylidene octyloxyaniline; $\bar{8}$S5, octyloxypentylphenyl thiolbenzoate.

How should we view this situation? It is widely felt that the problem is complicated by the Landau–Peierls instability or lack of true long-range order (Als-Nielsen *et al.* 1980) in the SmA phase. Some theorists also feel (Lubensky 1982) that the liquid crystal 'gauge' causes problems, because not all physical properties of the liquid crystal are gauge-invariant. (In superconductors one has $\nabla \cdot \hat{A} = 0$, while the splay elastic constant K_1 means $\nabla \cdot \hat{n} \neq 0$ in the liquid crystal gauge.) The dislocation models not only predict $\nu_\parallel = 2\nu_\perp$, but also that the elastic constants B and K_2 are constant at the N–SmA transition. Most experiments (Von Känel & Litster 1981; Litster *et al.* 1979 and references therein) show that B vanishes approximately as $t^{0.3}$; however, recent work (Fisch *et al.* 1982) on mixtures is consistent with a finite value of B at T_{NA}. It is hard to measure K_2 because of background masking the divergence, but early work (Delaye *et al.* 1973; Chu & McMillan 1975) suggests either critical or mean-field-like divergence. In my view the situation remains somewhat clouded experimentally. We need better measurements of K_2 and the elastic constants B and D. The behaviour of $\xi_\parallel$ and $\xi_\perp$ (table 1) seems well established; however, I should like to caution in its interpretation. Studies of re-entrant behaviour (Kortan *et al.* 1981) of the N–SmA transition in mixtures show that coupling of the smectic order parameter to other quantities (e.g. a temperature-dependent smectic interaction through thermal density changes) can change the divergences observed. Thus the exponents of table 1 may not be intrinsic to the N–SmA transition. There is a trend apparent in table 1 with the ratio of the N–SmA transition temperature to the nematic–isotropic transition temperature (T_{NI}). It is well known empirically that the N–SmA transition temperature becomes first-order when T_{NA}/T_{NI} is large enough. It is therefore possible that the smallest exponents of table 1 are influenced by the presence of a tricritical point (Brisbin *et al.* 1979). Experiments to clarify the situation are in progress in several laboratories around the world, and the experimental situation may be clarified within the year. Theoretically, in my view, the problem is still not solved.

(b) The smectic A to smectic C transition

When many liquid crystals in a SmA phase are cooled they have a second-order transition to a SmC phase. In the model of de Gennes (1974) the order parameter has two degrees of freedom: these are the tilt angle between the director and the smectic density wave, and the azimuthal angle of the director in the plane normal to the density wave vector. This transition should be in the same universality class as the superfluid transition in helium. The SmA–SmC transition may be studied by the usual methods: high-resolution X-ray scattering, light scattering, optical measurements, and precision calorimetry.

It has been studied by these methods by a number of laboratories and until recently the

[79]

situation had been somewhat confused. The SmC order parameter is the tilt Φ of the director with respect to the SmC density wave. It may be directly measured by holding the director fixed in a magnetic field and measuring the angle between the field and the directions for X-ray scattering from the density wave. Experiments by Safinya et al. (1980) showed that in the SmC phase Φ vanished as $|t|^{0.47\pm0.04}$, as would be expected from mean-field behaviour. This was contrary to the $|t|^{0.35}$ prediction of the 3D x–y model and the helium analogue; Safinya et al. argued that the bare length characterizing the SmA–SmC transition was large and that the Ginzburg (1960) criterion made the critical region too small to observe in the material, octyloxy-pentylphenylthiolbenzoate ($\overline{8}$S5), that they had studied. Subsequently Galerne (1981) reported that by optical measurements $\Phi \approx |t|^{0.36}$ in the SmC phase of azoxy-4,4′-di-undecyl-α-methyl-cinnamate (AMC-11) and argued that the bare length was anomalously small. Although the Ginzburg criterion need not require crossover from mean-field to critical behaviour at identical reduced temperatures in all materials, the situation seemed unsatisfactory. Huang & Viner (1982) made the important observation that the specific heat anomaly at the SmA–SmC transition in 4-(2′-methylbutyl)phenyl 4′-n-nonyloxybiphenyl-4-carboxylate (2M4P9OBC) could be explained by a mean-field model with an unusually large sixth-order term in the Landau expansion of the free energy. To clarify the situation we carried out a series of experiments at M.I.T. on 40.7. These showed (Birgeneau et al. 1983) that the order parameter could be described by a $|t|^{0.37}$ power law in the SmC phase, consistent with Galerne's observations. They also showed that one must question the uniqueness of power-law fits to data, for light scattering from fluctuations in short-range SmC order in the SmA phase showed quite clearly a susceptibility diverging as t^{-1}, unambiguously confirming mean-field behaviour. To understand the order parameter behaviour let us consider the mean-field model used by Huang & Viner. Near the SmA–SmC transition, the free energy is written

$$F = at\Phi^2 + b\Phi^4 + c\Phi^6 + \ldots + (1/2M_{\parallel})\,|\nabla_{\parallel}\Phi|^2 + (1/2M_{\perp})\,|\nabla_{\perp}\Phi|^2 + \ldots. \tag{11}$$

If one puts $t_0 = b^2/ac$, standard thermodynamic calculations yield

$$\Phi = (b/3c)^{\frac{1}{2}}\{(1 - 3t/t_0) - 1\}^{\frac{1}{2}} \quad (t < 0) \tag{12}$$

and a heat capacity

$$c = 0, \quad t > 0$$

$$= (a^2T/2bT_c^2)\,(1 - 3t/t_0)^{-\frac{1}{2}} \quad (t < 0). \tag{13}$$

Now, (12) predicts $|t|^{0.25}$ behaviour for $|t| \gg t_0$ and crosses over to $|t|^{0.50}$ for $|t| \ll t_0$. We found it fitted the data very well with a crossover temperature $t_0 = 1.3 \times 10^{-3}$ and could not be distinguished from power-law behaviour over the measured range of $-2.5 \times 10^{-3} < t < -1.0 \times 10^{-4}$. The data and fit to (12) are shown in figure 3. In figure 3 the light scattering data in the SmA phase, $T > T_c$, are shown. From (11) one predicts $\xi_{\parallel} = (2M_{\parallel}at)^{-\frac{1}{2}} = \xi_{\parallel}^0 t^{-\frac{1}{2}}$, $\xi_{\perp} = (2M_{\perp}at)^{-\frac{1}{2}} = \xi_{\perp}^0 t^{-\frac{1}{2}}$, and $\chi = \frac{1}{2}a_0 t = \chi_0 t^{-1}$; by using these the reciprocal of the intensity of scattered light, I^{-1}, may be expressed as

$$e_a^2 kT\chi_0 I^{-1} = t + (\xi_{\parallel}^0 q_z)^2 + (\xi_{\perp}^0 q_{\perp})^2, \tag{14}$$

where ϵ_a is the anisotropy in the dielectric constant of the SmA phase. The linear asymptotic behaviour of I^{-1} in figure 3 clearly confirms the mean-field behaviour, and the slope and value at T_c determine $\xi_{\parallel}^0 = 20.4 \pm 0.7$ Å†; the bare length $\xi_{\perp}^0$ was similar in magnitude but less accurately determined. Birgeneau et al. (1983) also report specific heat data for the SmA–SmC transition

† 1 Å $= 10^{-10}$ m $= 10^{-1}$ nm.

in 40.7; these are shown in figure 4. The SmA–SmC transition lies close to a SmB phase that contributes a background. This was determined by measurements of 40.8, which lacks the SmA–SmC transition. The fit to the background term plus equation (13) is shown in figure 4; it is excellent and yields the crossover temperature $t_0 = 1.3 \times 10^{-3}$ and heat capacity jump $\Delta C = 1.8 \times 10^5 \, \mathrm{J \, m^{-3} \, K^{-1}}$ at T_c. Combining all the data we calculate the Ginzburg reduced temperature for crossover to critical behaviour,

$$t_c = k_B^2 \left[32\pi^2 (\xi_{\parallel}^0)^2 \, (\xi_{\perp}^0)^4 \, (\Delta C)^2 \right]^{-1}, \tag{15}$$

to be $t_c = 3 \times 10^{-7}$, which explains why only mean-field behaviour is observed. Birgeneau et al. also made specific heat measurements on AMC-11 that show the same behaviour to obtain in that material. It seems likely that all SmA–SmC transitions are mean-field because of the long correlation lengths, quite analogous to the situation in superconductors. There is a lesson to be learned: that power law fits are not always unique and that several properties must be measured together to have a correct picture of behaviour at a phase transition. There is one material, p-nonyloxybenzoate-p-butyloxyphenol, for which light-scattering measurements (Delaye 1979) self-consistently show critical behaviour; it would be valuable to have high-resolution heat-capacity and X-ray measurements for this material.

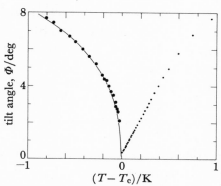

FIGURE 3. Large circles are the tilt order parameter, Φ, in the SmC phase of 40.7. The solid line is a fit to equation (12). Small circles are the reciprocal of the intensity (arbitrary units) scattered by tilt fluctuations in the SmA phase of 40.7. (From Birgeneau et al. (1983).)

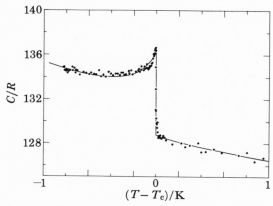

FIGURE 4. The specific heat in units of R near the SmA–SmC transition in 40.7. The solid line is a fit to equation (13) plus background, as discussed in the text. (From Birgeneau et al. (1983).)

152   J. D. LITSTER

(c) Lyotropic liquid crystal phases

As I discussed earlier, we believe that the fundamental symmetries of an ordered phase determine its phase transition behaviour; the microscopic details of the interactions are frequently unimportant. There are lyotropic materials that are quite different microscopically from the thermotropic nematics and smectics, yet have the same gross symmetry. Thus it is interesting to explore in detail the phase transition behaviour of these lyotropic phases. At M.I.T. we have recently begun experiments (Kumar *et al.* 1982) on a system that we were introduced to by A. Saupe (Haven *et al.* 1981). This is the soap decylammonium chloride (DACl) in aqueous solution with NH_4Cl. The DACl forms micelles that, with the appropriate NH_4Cl concentration, are disc-shaped. When the micelles occupy about 50 % of the total volume, interaction between them leads to collective behaviour. An isotropic solution of micelles has a first-order phase change to a nematic in which the micellar symmetry axes tend to orient parallel to the director. Further cooling of the solution leads to a second-order transition to a neat soap phase; this transition ought to be in the same universality class as thermotropic SmA–nematic transitions. We have found the nematic–isotropic transition to behave quite analogously to that of thermotropic materials (Stinson *et al.* 1972). The nematic elastic constants, as measured by light scattering from director modes show a pretransitional divergence for K_2 and K_3 as $|t|^{-0.68\pm0.05}$. This is consistent with the 3D x–y model, and preliminary results show no sign of the anisotropy in critical exponents commonly observed in thermotropic SmA phases. It will be interesting to learn more about these interesting lyotropic phases as experimental studies proceed.

3. Conclusions

We see that the nematic–SmA transition remains a problem, as it has for several years now. It seems likely that sufficient data will be accumulated within a year or so to provide a phenomenological description of this interesting transition. It is less certain when the theoretical difficulties will be overcome. The SmA–SmC situation has recently been clarified. A mean-field description with important sixth-order terms suffices for most materials because of the long bare lengths characteristic of the problem; a microscopic understanding of this remains to be found. It appears that a number of lyotropic phases may fruitfully be examined by the approaches used for thermotropic phases, and interesting results should appear in the future.

The M.I.T. work that I have described is the result of a stimulating collaboration with my colleagues Bob Birgeneau and Carl Garland as well as several postdoctoral associates and graduate students. It is a pleasure to express my appreciation to them. My participation in that work was supported by the National Science Foundation either under grant no. DMR 78-23555 or grant no. DMR 81-19295.

REFERENCES

Als-Nielsen, J., Litster, J. D., Birgeneau, R. J., Kaplan, M., Safinya, C. R., Lindegaard-Andersen, A. & Mathiesen, B. 1980 *Phys. Rev.* B **22**, 312.
Birgeneau, R. J., Garland, C. W., Kasting, G. B. & Ocko, B. M. 1981 *Phys. Rev.* A **24**, 2624.
Birgeneau, R. J., Garland, C. W., Kortan, A. R., Litster, J. D., Meichle, M., Ocko, B. M., Rosenblatt, C. & Yu, L.-J. 1983 *Phys. Rev.* A. (In the press.)
Brisbin, D., De Hoff, R., Lockhart, T. E. & Johnson, D. L. 1979 *Phys. Rev. Lett.* **43**, 1176.
Chen, J. H. & Lubensky, T. C. 1978 *Phys. Rev.* B **17**, 366.

Chu, K. C. & McMillan, W. L. 1975 *Phys. Rev.* A **11**, 1059.
Delaye, M. 1979 *J. Phys., Paris* **40**, C3-350.
Delaye, M., Rebotta, R. & Durand, G. 1973 *Phys. Rev. Lett.* **31**, 443.
Fisch, M. R., Sorensen, L. B. & Pershan, P. S. 1982 *Phys. Rev. Lett.* **48**, 943.
Frank, F. C. 1958 *Disc. Faraday Soc.* **25**, 19.
Galerne, Y. 1981 *Phys. Rev.* A **24**, 2284.
de Gennes, P. G. 1972 *Solid St. Commun.* **10**, 753.
de Gennes, P. G. 1974 *The physics of liquid crystals.* London: Oxford University Press.
Ginzburg, V. L. 1960 *Soviet Phys. solid St.* **2**, 1824.
Ginzburg, V. L. & Landau, L. D. 1950 *Soviet Phys. JETP* **20**, 1064.
Haven, T., Armitage, D. & Saupe, A. 1981 *J. chem. Phys.* **75**, 352.
Helfrich, W. 1978 *J. Phys., Paris* **39**, 1199.
Huang, C. C. & Viner, J. M. 1982 *Phys. Rev.* A **25**, 3385.
Huberman, B. A., Lubkin, D. M. & Doniach, S. 1975 *Solid St. Commun.* **17**, 485.
Jähnig, F. & Brochard, F. 1974 *J. Phys., Paris* **35**, 301.
Kortan, A. R., Von Känel, H., Birgeneau, R. J. & Litster, J. D. 1981 *Phys. Rev. Lett.* **47**, 1206.
Kumar, S., Sprunt, S. N., Yu, L.-J. & Litster, J. D. 1982 *Bull. Am. phys. Soc.* **27**, 355.
Litster, J. D., Als-Nielsen, J., Birgeneau, R. J., Dana, S. S., Davidov, D., Garcia-Golding, F., Kaplan, M., Safinya, C. R. & Shaetzing, R. 1979 *J. Phys., Paris* **20**, C3-339.
Lubensky, T. C. 1982 *J. Phys. appl.* (In the press.)
McMillan, W. L. 1971 *Phys. Rev.* A **4**, 1238.
Nelson, D. R. & Toner, J. 1981 *Phys. Rev.* B **23**, 363.
Safinya, C. R., Kaplan, M., Als-Nielsen, J., Birgeneau, R. J., Davidov, D. & Litster, J. D. 1980 *Phys. Rev.* B **21**, 4149.
Stinson, T. W., Litster, J. D. & Clark, N. A. 1972 *J. Phys., Paris* **33**, C1-69.
Von Känel, H. & Litster, J. D. 1981 *Phys. Rev.* A **23**, 3251.

Phil. Trans. R. Soc. Lond. A **309**, 155–165 (1983)
Printed in Great Britain

Some topics in continuum theory of nematics

By F. M. Leslie

Department of Mathematics, University of Strathclyde, Livingstone Tower,
26 Richmond Street, Glasgow G1 1XH, U.K.

This paper presents a concise formulation of continuum theory for nematic liquid crystals, both static and dynamic theory being discussed in turn. The emphasis is on the various assumptions behind the theory, and the experimental evidence supporting them. The resulting theory contains a fair number of material parameters, and progress concerning their measurement is assessed. Also, whenever possible, mention is made of problems that remain unresolved.

1. Introduction

My primary aim in this paper is to present a concise but clear derivation of continuum equations commonly employed to describe static and dynamic phenomena in nematic liquid crystals. My starting point is a formulation of static theory due to Ericksen (1962), which clarifies mechanical aspects and leads naturally to dynamic theory as a logical development of the well established static theory. To keep the presentation reasonably brief attention is confined to mechanical behaviour, and thermal considerations are largely excluded. Whenever possible, my presentation attempts to bring out experimental evidence to motivate certain assumptions made in the theory, as well as to serve as a means of either justifying or rejecting the theory.

It is now fairly widely accepted that the continuum theory discussed in this paper describes the mechanical response of nematic liquid crystals reasonably well. This belief stems largely from a number of analyses based on its linearized form which satisfactorily interpret various experimental observations. For example, one can cite in this context light-scattering, flow-induced instabilities, thermal convection and electrohydrodynamic instabilities; accounts of these topics are readily available in the books by de Gennes (1974) and Chandrasekhar (1977), and the reviews by Stephen & Straley (1974), Jenkins (1978) and Leslie (1979). Here, however, my concern is to show that the theory has the correct *nonlinear* form, and therefore my attention is mainly given to predictions and related experimental evidence in support of this claim.

First attempts to measure the various material parameters in the theory led to a fair spread of values in many cases. There were of course certain reasons for this; for example some of the methods of measurement were not wholly satisfactory, or impurities may have been present in the materials studied, or some of the nematics were simply chemically unstable. These factors had the unfortunate effect of encouraging a greater tolerance of experimental error than might normally have been acceptable. However, recently, with more stable materials and better methods, improved results are emerging, and it is of interest to discuss such developments where appropriate.

To present the continuum equations concisely it is convenient to use Cartesian tensor notation from time to time so that repeated indices are subject to the summation convention, a comma followed by an index denotes partial differentiation with respect to the corresponding spatial coordinate, and a superposed dot indicates a material time derivative.

2. Static theory

Formulation of the theory

Oseen (1925) was the first to propose a continuum theory to model equilibrium configurations of liquid crystals. He employed a unit vector field or director $\boldsymbol{n}(\boldsymbol{x})$ to describe the alignment of the anisotropic axis in these transversely isotropic liquids, and assumed that spatial distortions of this axis give rise to an energy density of the form

$$W = W(\boldsymbol{n}, \nabla \boldsymbol{n}), \tag{2.1}$$

quadratic in the gradients of the vector. His theory was later reappraised by Frank (1958), who gave an improved derivation of the energy function, essentially assuming invariance to superposed rigid rotations and to the sense of the director $\boldsymbol{n}$, so that

$$W(\boldsymbol{n}, \nabla \boldsymbol{n}) = W(\boldsymbol{Q}\boldsymbol{n}, \boldsymbol{Q}\nabla \boldsymbol{n}\,\boldsymbol{Q}^T) = W(-\boldsymbol{n}, -\nabla \boldsymbol{n}), \tag{2.2}$$

the second-order tensor $\boldsymbol{Q}$ being proper orthogonal. In nematics, material symmetry further restricts the functional dependence by the inclusion of improper transformations in the above condition. Hence, if the dependence upon gradients is at most quadratic, one finds for nematics that

$$2W = k_{11}(\operatorname{div}\boldsymbol{n})^2 + k_{22}(\boldsymbol{n}\cdot\operatorname{curl}\boldsymbol{n})^2 + k_{33}(\boldsymbol{n}\times\operatorname{curl}\boldsymbol{n})\cdot(\boldsymbol{n}\times\operatorname{curl}\boldsymbol{n})$$
$$+ (k_{22}+k_{24})\operatorname{div}\{(\boldsymbol{n}\cdot\operatorname{grad})\boldsymbol{n} - (\operatorname{div}\boldsymbol{n})\boldsymbol{n}\}, \tag{2.3}$$

the coefficients being the familiar Frank constants, which are at most functions of temperature. One approach commonly followed obtains equilibrium configurations by considering the Euler–Lagrange equations for the minimization of this energy, a further contribution being included to describe the influence of external magnetic or electric fields if present. The boundary conditions generally adopted assume that the solid surfaces confining the liquid crystal dictate a particular known orientation to the anisotropic axis at the surfaces, this depending upon their prior treatment.

It is of interest to interpret the above in mechanical terms, this being helpful for the development of a theory to describe flow and other dynamic effects. To this end, Ericksen (1961) postulates a principle of virtual work of the form

$$\delta \int_V W \, \mathrm{d}v = \int_V (\boldsymbol{F}\cdot\delta\boldsymbol{x} + \boldsymbol{G}\cdot\Delta\boldsymbol{n})\,\mathrm{d}v + \int_{\partial V}(\boldsymbol{t}\cdot\delta\boldsymbol{x} + \boldsymbol{s}\cdot\Delta\boldsymbol{n})\,\mathrm{d}a, \tag{2.4}$$

where
$$\Delta\boldsymbol{n} = \delta\boldsymbol{n} + (\delta\boldsymbol{x}\cdot\operatorname{grad})\boldsymbol{n}. \tag{2.5}$$

Here V denotes a volume of liquid crystal with surface ∂V, $\boldsymbol{F}$ and $\boldsymbol{t}$ are body and surface forces, and $\boldsymbol{G}$ and $\boldsymbol{s}$ generalized body and surface forces, respectively. Since incompressibility is assumed and $\boldsymbol{n}$ remains a unit vector, the variation in displacement $\delta\boldsymbol{x}$ and the material variation in the director $\Delta\boldsymbol{n}$ are subject to the constraints

$$\operatorname{div}\delta\boldsymbol{x} = 0, \quad \boldsymbol{n}\cdot\Delta\boldsymbol{n} = 0. \tag{2.6}$$

As Ericksen (1970) has shown, consideration of an arbitrary, infinitesimal, rigid translation in which $\Delta\boldsymbol{n}$ is zero leads to the somewhat obvious balance of forces

$$\int_V \boldsymbol{F}\,\mathrm{d}v + \int_{\partial V}\boldsymbol{t}\,\mathrm{d}a = 0. \tag{2.7}$$

[86]

However, similar consideration of an arbitrary, infinitesimal, rigid rotation $\boldsymbol{\omega}$, in which

$$\delta \boldsymbol{x} = \boldsymbol{\omega} \times \boldsymbol{x}, \quad \Delta \boldsymbol{n} = \boldsymbol{\omega} \times \boldsymbol{n}, \tag{2.8}$$

leads to a balance of couples

$$\int_V (\boldsymbol{x} \times \boldsymbol{F} + \boldsymbol{n} \times \boldsymbol{G})\, \mathrm{d}v + \int_{\partial V} (\boldsymbol{x} \times \boldsymbol{t} + \boldsymbol{n} \times \boldsymbol{s})\, \mathrm{d}a = 0. \tag{2.9}$$

Consequently the generalized forces are related to a body couple $\boldsymbol{K}$ and a couple stress $\boldsymbol{l}$ through

$$\boldsymbol{K} = \boldsymbol{n} \times \boldsymbol{G}, \quad \boldsymbol{l} = \boldsymbol{n} \times \boldsymbol{s}. \tag{2.10}$$

With the usual tetrahedron argument, (2.7) and (2.9) become in point form

$$t_i = t_{ij} v_j, \quad F_i + t_{ij,j} = 0, \tag{2.11}$$

and

$$l_i = l_{ij} v_j, \quad K_i + e_{ijk} t_{kj} + l_{ij,j} = 0, \tag{2.12}$$

respectively, $\boldsymbol{v}$ denoting the unit surface normal and e_{ijk} being the alternating tensor. Also, returning to (2.4) one finds that

$$t_{ij} = -p\delta_{ij} - (\partial W/\partial n_{k,j})\, n_{k,i}, \quad l_{ij} = e_{ipk} n_p\, \partial W/\partial n_{k,j}, \tag{2.13}$$

where p is an arbitrary pressure arising from the assumed incompressibility, and δ_{ij} is the familiar Kronecker delta. These latter results entail some manipulation of the left-hand side of (2.4), and details are available in Ericksen's paper.

Alternatively, however, if one rewrites the couple stress tensor l_{ij} as

$$l_{ij} = e_{ipk} n_p s_{kj}, \quad s_{ij} = n_i \beta_j + \partial W/\partial n_{i,j}, \tag{2.14}$$

where $\boldsymbol{\beta}$ is an arbitrary vector, (2.12) with (2.10) becomes

$$e_{ijk} n_j (G_k + s_{kp,p}) + e_{ijk}(t_{kj} + s_{kp} n_{j,p}) = 0. \tag{2.15}$$

Hence it is possible to introduce a vector $\boldsymbol{g}$ defined by

$$e_{ijk} n_j g_k = e_{ijk}(t_{kj} + s_{kp} n_{j,p}), \tag{2.16}$$

and the balance of couples (2.9) takes the alternative point form

$$G_i + g_i + s_{ij,j} = 0, \tag{2.17}$$

since from its definition $\boldsymbol{g}$ is indeterminate to the extent of an arbitrary scalar multiple of the director $\boldsymbol{n}$. Moreover, the invariance assumption (2.2) yields the following identity

$$e_{ijk}(n_j \partial W/\partial n_k + n_{j,p} \partial W/\partial n_{k,p} + n_{p,j} \partial W/\partial n_{p,k}) = 0, \tag{2.18}$$

first noted by Ericksen (1961). This, combined with (2.13), (2.14) and the definition of the intrinsic body force $\boldsymbol{g}$, gives

$$g_i = \gamma n_i - (n_i \beta_j)_{,j} - \partial W/\partial n_i, \tag{2.19}$$

with γ an arbitrary multiplier.

As for example Ericksen (1962) discusses, the body force and couple arising from an external magnetic or electric field can be expressed in the form

$$F_i = \partial \psi/\partial x_i, \quad G_i = \partial \psi/\partial n_i, \tag{2.20}$$

where for a magnetic field $\boldsymbol{H}$

$$2\psi = 2\psi_{\mathrm{m}} = \chi_\perp \boldsymbol{H} \cdot \boldsymbol{H} + \chi_{\mathrm{a}}(\boldsymbol{n} \cdot \boldsymbol{H})^2, \tag{2.21}$$

[87]

and for an electric field E

$$2\psi = 2\psi_{\mathrm{e}} = \epsilon_{\perp} \boldsymbol{E} \cdot \boldsymbol{E} + \epsilon_{\mathrm{a}} (\boldsymbol{n} \cdot \boldsymbol{E})^2, \tag{2.22}$$

the coefficients χ and ϵ being material parameters. This being so, Ericksen shows that the balance of forces (2.11) integrates with the aid of (2.13), (2.14), (2.17) and (2.19) to yield

$$p + W - \psi = p_0 \tag{2.23}$$

where p_0 is an arbitrary constant. Also, the balance of couples (2.17) becomes

$$(\partial W/\partial n_{i,j})_{,j} - \partial W/\partial n_i + \gamma n_i + G_i = 0, \tag{2.24}$$

the generalized body force having the respective forms

$$G_i = \partial \psi_{\mathrm{m}}/\partial n_i = \chi_{\mathrm{a}} n_j H_j H_i, \quad G_i = \partial \psi_{\mathrm{e}}/\partial n_i = \epsilon_{\mathrm{a}} n_j E_j E_i, \tag{2.25}$$

should a magnetic or electric field be present. Clearly, therefore, Ericksen's formulation identifies the Euler–Lagrange equation mentioned above with a local balance of couples in the liquid crystal.

Freedericksz transitions and Frank constants

The first experiments to be candidates for analysis in terms of the above theory were conducted by Freedericksz & Zolina (1933). These experiments require that one first uniformly align a thin layer of nematic liquid crystal between parallel plates by suitable surface treatments, and then apply a magnetic (or electric) field perpendicular to this initial alignment. For small field strengths, the influence of the surfaces dominates, but as the field strength increases a critical value is reached beyond which the field becomes the stronger influence and distortion occurs. In their experiments, Freedericksz & Zolina found that the critical field strength varies inversely with distance between the plates for a given nematic. Other variations of the experiment are possible, but in all cases that use parallel plates the critical field strength is apparently inversely proportional to the layer thickness for a given material (see, for example, Ericksen (1976) or Deuling (1978)).

For such problems the above theory reduces to an analysis of solutions of (2.24) subject to some prescribed alignment at the boundaries, with of course the function W given by (2.3). Rather clearly, if one introduces dimensionless variables by

$$\boldsymbol{x} = l\boldsymbol{x}^*, \quad \boldsymbol{H} = H\boldsymbol{h}, \tag{2.26}$$

and suitably scales the multiplier γ, the field strength H, assumed constant, and the gap width l enter the problem solely through the product Hl, and therefore the predicted scaling necessarily agrees with that observed. Thus this simple exercise supports the choice of the quadratic energy (2.3), since presumably any additional nonlinearity in this function must lead to some predictions at odds with observations. Equally, the choice of boundary conditions and body couple is apparently vindicated.

Detailed calculations lead to precise predictions of the critical field strengths, these naturally involving the coefficients in the energy function (2.3). Not surprisingly, this led initially to attempts to evaluate these material parameters from experimental measurements of the critical field strengths. However, determination of the precise point at which distortion of the alignment commences is not an easy matter, and extrapolations from data above the thresholds can present difficulties, as Tough & Raynes (1979) discuss. Consequently, more recent attempts to evaluate these material parameters use measurements over a wide range above the critical fields, and determine values of the coefficients that provide a best fit to the experimental data. Such studies on capacitance measurements by Bradshaw *et al.* (1981) and on optical data by Scheuble *et al.* (1981) and Bunning *et al.* (1981) appear to have led to more accurate and consistent values for the coefficients in the energy function.

One exception to the above remarks is of course the fourth energy coefficient k_{24}. Since the corresponding term does not contribute to (2.24), this coefficient does not appear in any of the predicted critical field strengths, nor in the description of behaviour above threshold. It does, however, appear in the couple stress and the energy density in certain cases, but at present no method appears to be available for its measurement.

3. Dynamic theory

Formulation of theory

Given that static theory describes equilibrium configurations rather well, one naturally seeks to generalize it when proposing a theory for dynamic effects in nematic liquid crystals. One such approach is to replace the principle of virtual work by a corresponding energy balance and derive equations for conservation of linear and angular momenta from it by invariance arguments, as for example Leslie (1979) describes. This leads to the rather natural extensions of (2.11) and (2.12)

$$\rho \dot{v}_i = F_i + t_{ij,j} \tag{3.1}$$

and

$$\sigma \dot{\omega}_i = K_i + e_{ijk} t_{kj} + l_{ij,j}, \tag{3.2}$$

where $\boldsymbol{v}$ denotes velocity and $\boldsymbol{\omega}$ local angular velocity given by

$$\boldsymbol{\omega} = \boldsymbol{n} \times \dot{\boldsymbol{n}}, \tag{3.3}$$

any component parallel to $\boldsymbol{n}$ itself being immaterial in a director description. On account of the assumption of incompressibility, the velocity $\boldsymbol{v}$ is subject to the constraint

$$\operatorname{div} \boldsymbol{v} = 0, \tag{3.4}$$

and the density ρ is simply a constant. The coefficient σ is a constant inertial coefficient, but the inertial term in angular momentum is frequently neglected, being generally considered unimportant. As before, (3.2) can be recast as

$$\sigma \ddot{n}_i = G_i + g_i + s_{ij,j}, \tag{3.5}$$

where

$$e_{ijk} n_j g_k = e_{ijk} (t_{kj} + s_{kp} n_{j,p}), \tag{3.6}$$

this again by using (2.10) and (2.14).

A natural starting point in the search for appropriate constitutive relations is simply to add dynamic contributions to the relations (2.13), (2.14) and (2.19) of static theory, and therefore one writes

$$\left.\begin{aligned}
t_{ij} &= -p\delta_{ij} - (\partial W/\partial n_{k,j})\, n_{k,i} + \tilde{t}_{ij}, \\
s_{ij} &= n_i \beta_j + \partial W/\partial n_{i,j} + \tilde{s}_{ij}, \\
g_i &= \gamma n_i - (n_i \beta_j)_{,j} - \partial W/\partial n_i + \tilde{g}_i,
\end{aligned}\right\} \tag{3.7}$$

where from the definition (3.6) and the identity (2.18)

$$e_{ijk} n_j \tilde{g}_k = e_{ijk} (\tilde{t}_{kj} + \tilde{s}_{kp} n_{j,p}), \tag{3.8}$$

and these additional terms $\tilde{t}$, $\tilde{s}$ and $\tilde{g}$ all vanish in equilibrium. The early viscosity measurements by Miesowicz (1936) clearly indicate that the dynamic contribution to the stress tensor depends upon the alignment, and the somewhat related measurements by Zwetkoff (1939) suggest that it also depends upon the rate of change of the anisotropic axis. Hence for a relevant theory it seems necessary to assume that the dynamic contributions are all functions of the director, its material time derivative, and the velocity gradients. Initially it is not obvious that one should not include certain other variables in these relations, but those cited appear sufficient to cover many effects.

It seems reasonable to ask that the dynamic terms vanish in a rigid rotation, and this implies dependence upon the velocity gradients and the director velocity solely through the rate of strain tensor A and a vector N, where

$$2A_{ij} = v_{i,j} + v_{j,i}, \quad N_i = \dot{n}_i - W_{ik}n_k, \quad 2W_{ij} = v_{i,j} - v_{j,i}. \tag{3.9}$$

It is also natural to consider a linear dependence upon both, given the experimental evidence mentioned above. Further, because this theory does not distinguish between n and $-n$, the stress tensor must be unaffected by a change of sign in the vector n, but generalized stress and the intrinsic body force must change sign with it. Such arguments coupled with invariance to superposed rigid rotations and appeal to nematic symmetries lead to

$$\tilde{t}_{ij} = \alpha_1 n_k n_p A_{kp} n_i n_j + \alpha_2 N_i n_j + \alpha_3 N_j n_i + \alpha_4 A_{ij} + \alpha_5 A_{ik}n_k n_j + \alpha_6 A_{jk}n_k n_i, \tag{3.10}$$

$$\tilde{s}_{ij} = 0, \quad \tilde{g}_i = -\gamma_1 N_i - \gamma_2 A_{ik}n_k, \tag{3.11}$$

where

$$\gamma_1 = \alpha_3 - \alpha_2, \quad \gamma_2 = \alpha_6 - \alpha_5, \tag{3.12}$$

the coefficients being at most functions of temperature. Earlier derivations of these relations by Leslie (1968, 1979) show that the dynamic part of the generalized stress is zero by other arguments. Of course thermodynamic considerations place some restriction upon possible values for the coefficients in the above relations. Leslie (1968) argues that positive entropy production requires that

$$\tilde{t}_{ij}A_{ij} - \tilde{g}_i N_i \geqslant 0, \tag{3.13}$$

and Parodi (1970) adds one further restriction, namely

$$\gamma_2 = \alpha_3 + \alpha_2 \tag{3.14}$$

by appeal to an Onsager relation.

Scaling property

As Ericksen (1969) points out, the above equations have a rather surprising property when external body forces and couples are absent. If one scales spatial coordinates and time by

$$x = lx^*, \quad t = l^2 t^*, \tag{3.15}$$

where l is some constant, the transformed equations are identical to the original provided that one neglects the inertial term in (3.5) and selects

$$p = l^{-2}p^*, \quad \gamma = l^{-2}\gamma^*. \tag{3.16}$$

In this scaling, the director's being a unit vector is not altered, but velocity and stress transform according to

$$v^* = lv, \quad t^* = l^2 t. \tag{3.17}$$

This rather simple property leads quickly to certain conclusions that go a long way towards vindicating the special constitutive assumptions outlined above.

[90]

For example, consider a simple shear flow between parallel plates a distance l apart and subject to a relative shear velocity V. The above scaling at once transforms this given problem into simple shear flow between parallel plates with unit gap and relative shear velocity V^*, where

$$V^* = Vl, \tag{3.18}$$

but the prescribed surface alignments are unchanged. Given a unique solution to the problem, it follows that

$$n = n^* = n^*(x^*, V^*) = n^*(l^{-1}x, Vl), \tag{3.19}$$

because V^* is the sole parameter in the transformed problem. From this one notes that in simple shear an optical property can depend upon V only through the product Vl. Rather encouragingly, Wahl & Fischer (1973) find such a dependence in optical measurements on a flow that closely approximates to simple shear.

Moreover, if one defines an apparent viscosity η for simple shear by

$$\eta = \tau l/V, \tag{3.20}$$

where τ is the shear stress applied per unit area of plate, it follows that

$$\eta = \tau l^2/Vl = \tau^*/V^* = \eta^*. \tag{3.21}$$

Hence one finds that

$$\eta = \eta^* = \mathscr{F}(V^*) = \mathscr{F}(Vl), \tag{3.22}$$

$\mathscr{F}$ being some unknown function. Unfortunately no experimental data are available to allow verification or otherwise of this novel prediction.

However, a similar analysis is feasible for Poiseuille flow, and for this flow Atkin (1970) predicts that the flux Q per unit time from a capillary of radius R is related to the pressure gradient a by

$$Q = R\mathscr{K}(aR^3), \tag{3.23}$$

the function $\mathscr{K}$ being unknown. Moreover, if one defines as apparent viscosity in this instance by

$$\eta = \pi a R^4/8Q, \tag{3.24}$$

it immediately follows that

$$\eta = \mathscr{G}(Q/R), \tag{3.25}$$

the function $\mathscr{G}$ again being unknown. This result is in marked contrast with the corresponding one for an isotropic liquid where the dependence is upon Q/R^3, *but* it has been confirmed experimentally by Fisher & Fredrickson (1969). These authors initially plot the apparent viscosity as a function of the latter and obtain distinct curves for each of the four capillaries used. However, they proceed to show that their data neatly fit a single curve when plotted according to Atkin's prediction.

As Leslie (1979) argues, this experimental confirmation of Atkin's novel scaling, and also that by Wahl & Fischer for simple shear, are particularly important because they largely vindicate the particular form of the theory, especially the functional dependence assumed in the constitutive theory. Any generalization of the above constitutive relations must quickly lead to predictions at odds with these experimental observations. Also, as Leslie points out, given the rapid changes of alignment in the narrow capillaries employed by Fisher & Fredrickson, it is hard to imagine that additional terms omitted in the above derivation could be negligible in these experiments. In view of the relative importance of this topic, it is somewhat surprising that essentially only two experimental investigations examine it. However, any generalization or extension of the above theory for nematic liquid crystals seems premature until further experimental evidence points to violations of the scaling described above.

Flow alignment

Flow can exert a strong influence upon alignment in a nematic, and therefore a brief résumé of this topic is helpful before proceeding to a discussion of methods used to measure the various viscous coefficients. For most purposes it is sufficient to consider simple shear flow and to ignore complications arising from the competing influences of boundaries and external fields. Here, therefore, one examines solutions of the above equations in which the velocity and director have Cartesian components of the form

$$v_x = \kappa z, \quad v_y = v_z = 0, \quad \kappa > 0,$$
$$n_x = \cos\theta\cos\phi, \quad n_y = \cos\theta\sin\phi, \quad n_z = \sin\theta, \tag{3.26}$$

where κ, θ and ϕ are simply constants. This choice at once satisfies (3.1) and (3.4) rather trivially, and reduces (3.5) to

$$\kappa(\alpha_3\cos^2\theta - \alpha_2\sin^2\theta)\cos\phi = 0, \quad \kappa\alpha_2\sin\theta\sin\phi = 0, \tag{3.27}$$

the final expressions by using the relations (3.12) and (3.14). Whatever the values of α_2 and α_3, the above equations always have a solution

$$\theta = 0, \quad \phi = \tfrac{1}{2}\pi, \tag{3.28}$$

representing alignment normal to the plane of shear. In addition, if α_2 and α_3 have the same sign, there are two steady solutions in the plane of shear, namely

$$\theta = \pm\theta_0, \quad \phi = 0, \tag{3.29}$$

where the acute angle θ_0 is defined by

$$\tan^2\theta_0 = \alpha_3/\alpha_2. \tag{3.30}$$

The thermodynamic inequality requires that γ_1 be positive and so two possibilities emerge: either

$$\alpha_2 < \alpha_3 < 0 \quad (0 < \theta_0 < \tfrac{1}{4}\pi), \tag{3.31}$$

or

$$\alpha_3 > \alpha_2 > 0 \quad (\tfrac{1}{4}\pi < \theta_0 < \tfrac{1}{2}\pi). \tag{3.32}$$

For the frequently observed alignment at a small angle to the streamlines, the viscous coefficients must clearly satisfy the former.

An examination of time-dependent perturbations of the director shows that the solution (3.28) is always unstable, and this configuration and its instability have been studied extensively by Pieranski & Guyon (see, for example, Dubois-Violette *et al.* (1977)). If the solutions (3.29) are possible, one is stable to such perturbations, and the other unstable. When the coefficients satisfy (3.31) the stable configuration corresponds to that with the positive sign, but, if (3.32) applies, the stable solution has the negative sign. It is of course possible to measure this flow alignment angle θ_0 experimentally, and Gähwiller (1971, 1973) describes such measurements. While there are solutions of the equations that demonstrate the competition between flow and boundaries on alignment, their description is not attempted here, given their complexity. None the less, the response of such flow-aligning nematics in shear flow is now reasonably well understood (see, for example, Leslie (1981)).

However, as first discovered by Gähwiller (1972), there are nematics that do not align in shear flow, this being ascribed to the coefficient α_3 changing sign for at least part of its nematic temperature range, this generally preceding a transition to a smectic phase. For this type of nematic, the behaviour in shear flow is not as well understood. Two experimental studies, one by Pieranski *et al.* (1976) and the second by Cladis & Torza (1975), investigate this and some differences emerge. A variety of flow-induced instabilities certainly occur, possibly accompanied by other complications, but theoretical investigations appear to be limited to an approximate analysis by Pikin (1974), what is available in the two papers just cited, and a discussion of oscillatory shear by Clark *et al.* (1981). Certainly this topic deserves more attention than it has received, although progress may be difficult.

The behaviour in shear flow certainly typifies the response in other viscometric flows as far as one can tell at present. Little has been yet attempted with regard to flows that are not viscometric, presumably because there appears to have been little need for it. However, studies of other anisotropic systems tend to look to liquid crystal theory for guidance, and such interests may begin to motivate investigations of other flows.

Viscosity measurements

Viscosity measurements commonly employ a magnetic field to control alignment of the nematic, while measuring some quantity that yields a value for the corresponding viscosity. The assumption is that a sufficiently strong field completely aligns the liquid crystal, and the test of sufficient strength is generally that measurements appear to reach some limiting value. The theory employed is somewhat limited, simply assuming uniform alignment with possibly some assessment of likely errors due to variations at boundaries.

To examine the information obtainable from such measurements it suffices to consider simple shear flow with some arbitrary alignment (3.26). If one defines a viscosity by

$$\eta = t_{xz}/\kappa, \tag{3.33}$$

it follows straightforwardly that

$$\eta = \eta_c \sin^2\theta + (\eta_a \sin^2\phi + \eta_b \cos^2\phi)\cos^2\theta + \alpha_1 \sin^2\theta \cos^2\theta \cos^2\phi, \tag{3.34}$$

where

$$2\eta_a = \alpha_4, \quad 2\eta_b = \alpha_4 + \alpha_3 + \alpha_6, \quad 2\eta_c = \alpha_4 + \alpha_5 - \alpha_2. \tag{3.35}$$

Hence from such measurements one hopes to obtain three viscosities corresponding to particular alignments as follows:

$$\left. \begin{array}{llll} \text{(i)} & \theta = 0, & \phi = \tfrac{1}{2}\pi, & \eta = \eta_a, \\ \text{(ii)} & \theta = 0, & \phi = 0, & \eta = \eta_b, \\ \text{(iii)} & \theta = \tfrac{1}{2}\pi, & \phi = 0, & \eta = \eta_c, \end{array} \right\} \tag{3.36}$$

and the coefficient α_1 by a further measurement with alignment in the plane of shear.

Miesowicz (1936) was the first to attempt such viscosity measurements. He measured the damping of vertical oscillations of a plate suspended in a nematic, calibrating his measurements with a liquid of known viscosity. Because his magnetic field was horizontal he determined η_a and η_c. Subsequently, Gähwiller (1971, 1973) carried out a complete set of such measurements by measuring the flux through a rectangular capillary, again using a standard liquid for calibration. More recently Kneppe & Schneider (1981) and Skarp *et al.* (1980) have also used capillaries to make such measurements on nematics.

As de Jeu (1978) points out, there is some concern that the field strengths are sufficiently large to ensure complete alignment, particularly for η_c. Certainly many of the newer materials are more viscous than those first studied, and consequently greater fields are necessary to overcome the stronger flow alignment. A further cause for concern is the lack of any theory to guide extrapolation to limiting values. As Clark *et al.* (1980) show in a somewhat related situation, extrapolation without some theory to guide can be rather uncertain.

Given the above quantities accurately determined, the Parodi relation (3.14) and (3.12) yield

$$\eta_b - \eta_c = \gamma_2, \tag{3.37}$$

and hence γ_2 is also found. If the flow alignment angle is known, this provides the fifth measurement and the α are determined. However, if the nematic does not align in shear, some other measurement is necessary. One candidate is the experiment first employed by Zwetkoff (1939) in which he suspended some nematic in a cylinder and rotated a horizontal magnetic field about the axis of suspension. A measurement of the torque transmitted to the cylinder yields a value for γ_1. Here again the theory used is relatively simplistic, and details are given by Gasparoux & Prost (1971). More recently Gerber (1981) has described a similar experiment that appears to remove most of the problems associated with Zwetkoff's method.

At present insufficient measurements have been performed on a given material to allow meaningful comparisons and checks for consistency. It is to be hoped that the progress made with regard to the measurement of the elastic constants will encourage further efforts to determine the viscous constants more accurately. However, there is clearly a need to develop alternative methods, preferably with the use of small samples, although light scattering (see Orsay Liquid Crystal Group 1971) and shear wave reflectance (Martinoty & Candau 1971) are possible techniques meeting this requirement.

REFERENCES

Atkin, R. J. 1970 *Arch. ration. Mech. Analysis* **38**, 224–240.
Bradshaw, M. J., McDonnell, D. G. & Raynes, E. P. 1981 *Molec. Cryst. liq. Cryst.* **70**, 289–300.
Bunning, J. D., Faber, T. E. & Sherrell, P. L. 1981 *J. Phys., Paris* **42**, 1175–1182.
Chandrasekhar, S. 1977 *Liquid crystals*. Cambridge University Press.
Cladis, P. E. & Torza, S. 1975 *Phys. Rev. Lett.* **35**, 1283–1286.
Clark, M. G., Raynes, E. P., Smith, R. A. & Tough, R. J. A. 1980 *J. Phys.* D **13**, 2151–2164.
Clark, M. G., Saunders, F. C., Shanks, I. A. & Leslie, F. M. 1981 *Molec. Cryst. liq. Cryst.* **70**, 195–222.
de Gennes, P. G. 1974 *The physics of liquid crystals*. Oxford: Clarendon Press.
de Jeu, W. H. 1978 *Phys. Lett.* A **69**, 122–124.
Deuling, H. J. 1978 *Solid State Phys. Suppl.* **14**, 77–107.
Dubois-Violette, E., Guyon, E., Janossy, I., Pieranski, P. & Manneville, P. 1977 *J. Méc.* **16**, 733–767.
Ericksen, J. L. 1961 *Trans. Soc. Rheol.* **5**, 23–34.
Ericksen, J. L. 1962 *Arch. ration. Mech. Analysis* **9**, 371–378.
Ericksen, J. L. 1969 *Trans. Soc. Rheol.* **13**, 9–15.
Ericksen, J. L. 1970 General Lecture, British Theoretical Mechanics Colloquium, Norwich.
Ericksen, J. L. 1976 *Adv. liq. Cryst.* **2**, 233–298.
Fisher, J. & Fredrickson, A. G. 1969 *Molec. Cryst. liq. Cryst.* **8**, 267–284.
Frank, F. C. 1958 *Discuss. Faraday Soc.* **25**, 19–28.
Freedericksz, V. & Zolina, V. 1933 *Trans. Faraday Soc.* **29**, 919–930.
Gähwiller, C. 1971 *Phys. Lett.* A **36**, 311–312.
Gähwiller, C. 1972 *Phys. Rev. Lett.* **28**, 1554–1556.
Gähwiller, C. 1973 *Molec. Cryst. liq. Cryst* **20**, 301–318.
Gasparoux, H. & Prost, J. 1971 *J. Phys., Paris* **32**, 953–962.
Gerber, P. R. 1981 *Appl. Phys.* A **26**, 139–142.
Jenkins, J. T. 1978 *A. Rev. Fluid Mech.* **10**, 197–219.
Kneppe, H. & Schneider, F. 1981 *Molec. Cryst. liq. Cryst.* **65**, 23–38.
Leslie, F. M. 1968 *Arch. ration. Mech. Analysis* **28**, 265–283.
Leslie, F. M. 1979 *Adv. liq. Cryst.* **4**, 1–81.
Leslie, F. M. 1981 *Molec. Cryst. liq. Cryst.* **63**, 111–128.
Martinoty, P. & Candau, S. 1971 *Molec. Cryst. liq. Cryst.* **14**, 243–271.
Miesowicz, M. 1936 *Bull. int. Acad. pol. Sci. Lett.* A, pp. 228–247.

Orsay Liquid Crystal Group 1971 *Molec. Cryst. liq. Cryst.* **13**, 187–191.

Oseen, C. W. 1925 *Ark. Mat. Astr. Fys.* A **19**, 1–19.

Parodi, O. 1970 *J. Phys., Paris* **31**, 581–584.

Pieranski, P., Guyon, E. & Pikin, S. A. 1976 *J. Phys., Paris* **37**(C1), 3–6.

Pikin, S. A. 1974 *Soviet Phys. JETP* **38**, 1246–1250.

Scheuble, B. S., Baur, G. & Meier, G. 1981 *Molec. Cryst. liq. Cryst.* **68**, 57–67.

Skarp, K., Lagerwall, S. T. & Stebler, B. 1980 *Molec. Cryst. liq. Cryst.* **60**, 215–236.

•Stephen, M. J. & Straley, J. P. 1974 *Rev. mod. Phys.* **46**, 617–704.

Tough, R. J. A. & Raynes, E. P. 1979 *Molec. Cryst. liq. Cryst. Lett.* **56**, 19–25.

Wahl, J. & Fischer, F. 1973 *Molec. Cryst. liq. Cryst.* **22**, 359–373.

Zwetkoff, W. 1939 *Acta phys.-chim. URSS* **10**, 555–578.

Phil. Trans. R. Soc. Lond. A **309**, 167–178 (1983)

Printed in Great Britain

Electro-optic and thermo-optic effects in liquid crystals

By E. P. Raynes

*Royal Signals and Radar Establishment, St Andrews Road, Great Malvern,
Worcestershire WR14 3PS, U.K.*

The twisted nematic electro-optic effect has become widely used as a low-voltage, low-power display in watches and calculators; however, it is only one of the many optical effects found in the various liquid crystal phases. Despite this wide variety, certain features of operation are general. Liquid crystals are birefringent, which, together with their ability to align on solid surfaces, allows the construction of thin layers with optical properties reminiscent of single solid crystals. The more common display devices use anisotropic electrical properties to produce electro-optic effects that are seen by using polarized light, scattering, or the absorption of light. Thermo-optic effects can be produced by varying the temperature of the liquid crystal in the vicinity of a phase transition, and are used in thermometry, thermography, and display devices.

1. Introduction

In this paper the major electro-optic and thermo-optic effects in liquid crystals will be reviewed. Selection of effects has been necessary, and some have been dealt with more briefly than their importance may warrant. The first part of the paper examines the building blocks of electro-optic and thermo-optic effects: the anisotropic properties of liquid crystals, ways of observing changes in orientation and construction of thin-layer devices. The second part looks at the various electro-optic and thermo-optic effects in the different liquid crystal phases.

2. Anisotropic properties

The anisotropic physical properties (de Jeu 1980) make liquid crystals both fascinating and useful. I shall now examine the anisotropic physical properties that contribute to display operation (Clark *et al.* 1980).

(a) Refractive index

The most obvious feature (invariably used by the organic chemist to identify a liquid crystal phase) is the large anisotropy of refractive index, or birefringence, given by

$$\Delta n = n_{\parallel} - n_{\perp},$$

where $n_{\parallel}$ and $n_{\perp}$ are the refractive indices measured parallel and perpendicular to the nematic director $\boldsymbol{n}$; typically $\Delta n \approx 0.25$.

(b) Electric permittivity

The analogous low-frequency property, the electric permittivity, is also anisotropic:

$$\Delta \epsilon = \epsilon_{\parallel} - \epsilon_{\perp}.$$

Liquid crystal molecules are free to rotate in an applied field, and permanent dipole moments within the molecule contribute to, and indeed usually dominate, the permittivity anisotropy. Cyano groups are frequently used in liquid crystal molecules, and their large dipole moment produces strong anisotropy of the permittivity, giving typically $\epsilon_{\perp} \approx 5$ and $\epsilon_{\parallel} \approx 20$.

168 E. P. RAYNES

The anisotropy of electric permittivity has become extremely important in the last few years because it is the 'driving force' in most electro-optic effects. The electric contribution to the free energy density contains a term that depends on the angle between the director n and the electric field E, and is given approximately by

$$F_E = -\tfrac{1}{2}\epsilon_0 \Delta\epsilon\,(n \cdot E)^2.$$ (1)

In the absence of any other constraints the director will rotate to minimize this contribution to the free energy. There are two possibilities: the director orients parallel to the field for positive anisotropy $(\epsilon_\parallel > \epsilon_\perp)$, and perpendicular to the field for negative anisotropy $(\epsilon_\parallel < \epsilon_\perp)$. An interesting variation, which can be useful in certain devices (Raynes & Shanks 1974), is the reversal in sign of the permittivity anisotropy (and hence field-induced orientation) with increasing frequency. In liquid crystals the normal Debye relaxation found in isotropic liquids in the region of 100 MHz is severely reduced by the nematic potential to 1 MHz or less (Maier & Meier 1961), and in extreme cases to 1 kHz. This influence of the nematic potential on the relaxation process only applies to the contribution of longitudinal dipoles to $\epsilon_\parallel$, so that $\epsilon_\parallel$ can relax below $\epsilon_\perp$ and the anisotropy change sign.

(c) Orientational elasticity

In nematic liquid crystals the preferred orientation is with the director everywhere parallel. All other orientations have a free energy density given by the Frank–Oseen continuum theory (Frank 1958)

$$F_k = \tfrac{1}{2}k_{11}(\mathrm{div}\,n)^2 + \tfrac{1}{2}k_{22}(n \cdot \mathrm{curl}\,n)^2 + \tfrac{1}{2}k_{33}(n \times \mathrm{curl}\,n)^2.$$ (2)

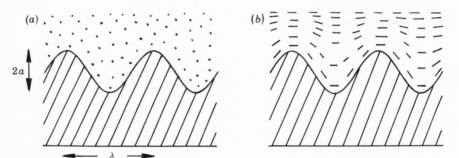

FIGURE 1. Alignment of a nematic liquid crystal on a microgrooved surface: (a) lowest-energy configuration with the director parallel to the microgrooves; (b) highest-energy configuration with the director perpendicular to the microgrooves.

The orientational elastic constants k_{11}, k_{22} and k_{33} describe the splay, twist and bend deformations respectively, and have a magnitude of 10^{-11} N. The orientation of liquid crystals on a microgrooved surface is an immediate consequence of (2). Surface alignment is usually a result of both the chemical nature and the topography of the interface of the liquid crystal and the surface. If the surface is microgrooved in one direction (figure 1) the difference in free energy between the two configurations is (Berreman 1972)

$$\Delta F_k = (\pi^3 k/2\lambda)\,(2a/\lambda)^2,$$ (3)

where k is an average elastic constant. With $k \approx 10^{-11}$ N, $2a/\lambda \approx 0.5$, and $\lambda \approx 0.05\,\mu$m, we find $\Delta F_k \approx 5 \times 10^{-4}\,\mathrm{J\,m^{-2}}$, the same order of magnitude as the energies associated with surface physico-chemical forces. A thin layer with two aligned surfaces adopts an orientation determined by the two surfaces and is the starting point for most optical effects in liquid crystals.

By adding the electric contribution F_E (equation (1)) to the orientation free energy F_k (equation (2)) it is possible to calculate many useful properties of electro-optic displays (Berreman, this symposium).

3. Visualization of orientational change

There are three main ways in which a change in the orientation of the liquid crystal can be seen by an observer. The changing refractive index can be observed by using two polarizers, one to polarize and one to analyse the light passing through the cell. Secondly, the change in anisotropic light absorption associated with the director reorientation can be observed directly or with a single polarizer. Most liquid crystals are transparent, so absorption is induced by adding pleochroic dye molecules (Heilmeier & Zanoni 1968). The last method of visualizing the director reorientation uses the scattering induced when the director becomes randomly oriented on a scale comparable with the wavelength of light. This scattering texture can be observed without polarizers against the clear aligned background areas.

4. Display construction

Display devices using the various electro-optic and thermo-optic effects have a fairly standard flat panel construction, shown in figure 2 (Clark *et al.* 1980). A thin layer of liquid crystal (*ca.* 10 μm) is sandwiched between two sheets of glass joined together by a thermoplastic or thermo-setting seal around the perimeter. Layer spacing is usually controlled by small pieces of glass

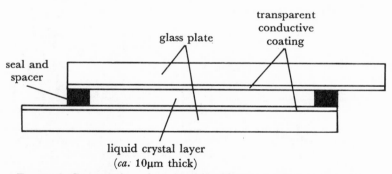

FIGURE 2. Construction of a typical liquid crystal cell (not to scale).

fibre of the appropriate diameter distributed across the display area. The liquid crystal is vacuum filled through a small gap left in the seal, which is subsequently closed by epoxy resin or solder. A field is applied across the liquid crystal by using inner glass surfaces coated with transparent conductors made of indium tin oxide and etched to provide the required pattern. Frequently a passivation layer is used to prevent damage by ion migration from the glass substrate and to act as a d.c. blocking layer to inhibit electrochemical degradation. Unidirectional homogeneous alignment of the liquid crystal is produced by using microgrooves produced either by a rubbing process or by obliquely evaporated dielectric layers (Janning 1972). Homeotropic alignment of the liquid crystal normal to the surface is achieved by chemical treatment of the surfaces with silane coupling agents (Kahn 1973*b*) or chrome complexes (Matsumoto *et al.* 1975).

5. Electro-optic effects

We have seen how a thin layer of liquid crystal can be aligned by the surfaces and reoriented by an electric field to produce a visual change in the optical properties. I shall now describe the various electro-optic effects that can be constructed out of these basic elements.

(a) Nematic liquid crystals

(i) Dynamic scattering

Dynamic scattering displays (Heilmeier *et al.* 1968) are worthy of a brief mention because although they are now rarely used, they were the first generation of liquid crystal displays. Their operation depends on a complex electrohydrodynamic effect (Helfrich 1969) not mentioned above; this disrupts the alignment to produce scattering of light, which is seen against a clear background of uniformly oriented liquid crystal. Dynamic scattering requires materials with $\Delta\epsilon < 0$, and in the early days such materials were unstable and contributed to a short lifetime for the displays. Although suitable stable materials now exist, the performance and use of dynamic scattering displays is limited by a number of factors:

short lifetime, resulting from the high electrical conductivity necessary for adequate scattering;
relatively high power consumption;
poor visual appearance (in reflexion a metallic or dielectric mirror was used to reflect the scattered light, resulting in specular reflexions in unenergized regions and a poor viewing angle and legibility).

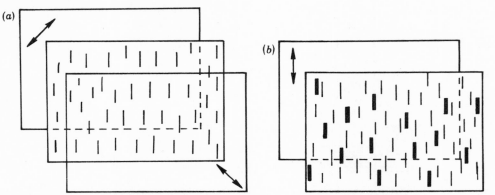

FIGURE 3. Two electro-optic effects in parallel aligned nematic layers: (*a*) variable birefringence with two polarizers; (*b*) guest–host effect with a single polarizer.

(ii) Variable birefringence displays

If a nematic liquid crystal with $\Delta\epsilon > 0$ is aligned unidirectionally in a thin layer by using two parallel aligned surfaces the change in birefringence when a field is applied can be observed by using two polarizers (figure 3*a*). The layer shows field-induced reorientation (Freedericksz & Zolina 1933) above a threshold voltage (*ca.* 1 V) given by

$$V_{\rm c} = \pi\sqrt{(k_{11}/\epsilon_0\Delta\epsilon)}. \tag{4}$$

It is also possible to use the opposite effect in a material with $\Delta\epsilon < 0$ in a homeotropically aligned layer. The variable birefringence effect (Schiekel & Fahrenschon 1971; Assouline 1971) has a performance limited by the strong dependence of contrast on wavelength, viewing angle, temperature and layer thickness. Its main use is a transmissive or projection display, particularly for multiplexed displays with a high information content.

(iii) *Single-polarizer guest–host effect*

The construction of this device is similar to the variable birefringence display (§5a(ii)) except that a pleochroic dye is dissolved in the liquid crystal and only one polarizer is used (figure 3b). The reorientation process is also identical with the threshold voltage given by (4), and the dye absorption decreases as the voltage is raised above V_c. The single-polarizer guest–host display (Heilmeier & Zanoni 1968) is well suited to transmissive or back-lit applications where its wide angle of view is an advantage. The effect just described shows negative contrast with bright characters on a dark background; the contrast is reversed if a material with $\Delta\epsilon < 0$ is used with homeotropic boundary conditions.

(iv) *Twisted nematic effect*

The vast majority of commercial liquid crystal displays use the twisted nematic electro-optic effect. The visualization of the twisted nematic effect uses a particularly subtle optical property not described in §3. The basic construction of the layer uses two aligned surfaces as in §§5a(ii) and (iii), but this time the grooving directions are orthogonal, so that the nematic director twists by 90° in going from one boundary surface to the other (figure 4). This twisted structure rotates

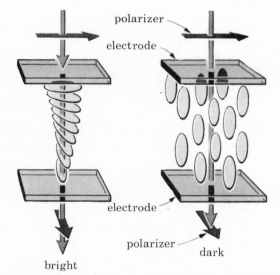

FIGURE 4. Operation of a conventional twisted nematic cell.

the plane of polarized light, which is guided by the twisted birefringent liquid crystal. This unusual property was first observed by Mauguin (1911) who also calculated that the guiding was effective only if

$$d\Delta n > 2\lambda, \tag{5}$$

where d is the thickness of the liquid crystal layer and λ the wavelength of the light. When an electric field is applied the director reorients above a threshold voltage (*ca.* 1 V) given by

$$V_c = \pi\sqrt{\{k_{11} + \tfrac{1}{4}(k_{33} - 2k_{22})/\epsilon_0 \Delta\epsilon\}}, \tag{6}$$

and the guiding property is lost (Schadt & Helfrich 1971). This can be understood quite simply from (5) as the director realigns along the electric field, the effective Δn is lowered below the level for guiding. The twisted nematic layer can be used as an electro-optic shutter or more commonly as a display device.

Normally, twisted nematic devices show degeneracy of twist and tilt producing an unacceptably patchy contrast. These problems are overcome by using very long-pitch cholesteric materials and particular combinations of twist and surface tilt (Raynes 1975). The finite pitch P modifies the threshold voltage (Raynes 1975),

$$V_c = \pi \sqrt{\frac{k_{11} + \frac{1}{4}(k_{33} - 2k_{22}) + 2k_{22}d/P}{\epsilon_0 \Delta\epsilon}}, \tag{7}$$

and slightly degrades the optical performance (Raynes 1979), producing an increased transmission (T) between crossed polarizers calculated by the author to be given approximately (for $V \gg V_c$) by

$$T \approx (2\pi d V_c / PV)^2. \tag{8}$$

The guiding is independent of wavelength above the limit set by (5) and therefore the displays are not coloured and have a high contrast ratio (more than 10:1). Good operation is achieved at low voltages ($ca.$ 2V) with a lower power consumption ($< 1\,\mu W\,cm^{-2}$). This performance, together with the long lifetimes now available, has contributed to making the twisted nematic display the leading low-voltage low-power display technology.

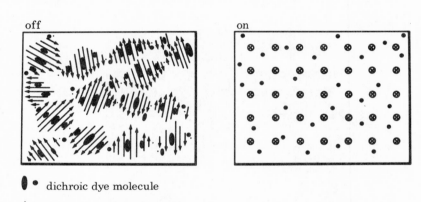

off   on

● • dichroic dye molecule

↑ ⊗ director of liquid crystal

FIGURE 5. Operation of a cholesteric–nematic phase-change cell with dye: (a) no applied field; (b) electric field applied perpendicular to plane of figure.

(b) Cholesteric liquid crystals

The major electro-optic effect in cholesteric liquid crystals uses materials with $\Delta\epsilon > 0$ that have pleochroic dyes dissolved in them. The natural pitch (P) of a cholesteric material is typically less than $10\,\mu m$, and the rate of twist of the director is therefore too rapid for guiding to occur (this can be seen by putting $d = \frac{1}{4}P$ in (5)), and consequently a dissolved pleochroic dye can absorb all polarizations of incident light (White & Taylor 1974). An electric field reorients the director, a process regarded as unwinding of the cholesteric helix in short-pitch materials (figure 5) and called the cholesteric–nematic phase-change effect (Wysocki $et\ al.$ 1968). The threshold field for this unwinding (Meyer 1968; de Gennes 1968) is given by

$$E_c = (\pi^2/P) \sqrt{(k_{22}/\epsilon_0 \Delta\epsilon)}, \tag{9}$$

and for $P \approx 3\,\mu m$, $\Delta\epsilon \approx 10$, and $k_{22} \approx 10^{-11}\,N$ corresponds to a threshold voltage of 10V in a $10\,\mu m$ layer. In longer pitch systems the reorientation is more nematic-like in nature (Suzuki *et al.* 1981) with a threshold voltage given approximately by (7) and $V_c \approx 2$ V. In both cases adequate contrast can be achieved without polarizers and the brightness and viewing angle are significantly better than for twisted nematic displays.

The cholesteric displays just described show clear characters on a coloured background. The contrast can be reversed by a number of specialized display construction techniques or by using materials with $\Delta\epsilon < 0$.

(c) *Smectic liquid crystals*

Smectic liquid crystals are attracting increasing interest for electro-optic applications because they can be switched in quite short times and usually show long-term memory, making them useful in multiplexed displays of high complexity.

(i) *Smectic A scattering effects*

By applying a low-frequency voltage (*ca.* 100 V) to an aligned smectic A layer with high electrical conductivity, permanent scattering somewhat reminiscent of dynamic scattering in nematics can be induced in quite short times (10–100 ms) (Tani 1971; Steers & Mircea-Roussel 1976; Coates *et al.* 1978). With homeotropic boundary conditions and $\Delta\epsilon > 0$ this effect can be reversed by the application of a high-frequency signal (Coates *et al.* 1978). Figure 6 shows

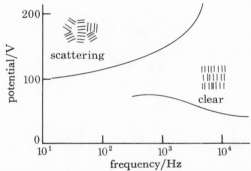

FIGURE 6. Threshold curves of a homeotropically aligned smectic A layer with $\Delta\epsilon > 0$.

typical threshold curves for both transitions. By applying a large voltage (*ca.* 100 V) and switching the frequency from 10^2 to 10^3 Hz a reversible transition from a clear to a scattering state is induced in times as short as 10 ms. Both states show long-term memory, making the effect useful in complex multiplexed displays (Crossland & Ayliffe 1982).

(ii) *Chiral smectic C ferroelectric effects*

The only known liquid crystal phase that can possess ferroelectric properties is the chiral smectic C phase (Meyer 1977), in which every smectic layer possesses an electric dipole density perpendicular to the director n and parallel to the smectic layer plane (figure 7a). Normally the director and polarization spiral along the helix (figure 7b), resulting in a cancellation of the polarization and a reduction of the bulk polarization to zero. However, this can be overcome by unwinding the helix. In the device described by Clark & Lagerwall (1980) the spiralling is suppressed by using a thin cell and a surface alignment that allows the director to lie down within the plane of the surface without imposing any specific direction. There are now two possible orientations of the director corresponding to the intersection of the cone of angle θ

174   E. P. RAYNES

(where θ is the tilt angle of the smectic C phase) with the surface. They are at an angle of 2θ to each other and have dipole moments pointing up and down respectively (figure 8a, b). The device is usually sheared just after cooling from the isotropic phase to produce uniform alignment. The director can then be switched back and forth through an angle of 2θ by altering the polarity of the applied electric field. By using crossed polarizers with the incident polarization direction parallel to the director in one of the polarization states there is extinction of light by the second polarizer for that state (figure 8c). The other polarization state transmits an intensity

$$I = I_0 \{\sin 4\theta \sin (\pi \Delta n d / \lambda)\}^2, \tag{10}$$

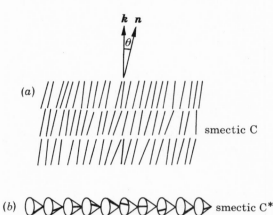

FIGURE 7. Smectic C configurations: (a) director tilted from layer normal $\textbf{\textit{k}}$ by θ; (b) spiralling of director in chiral smectic C.

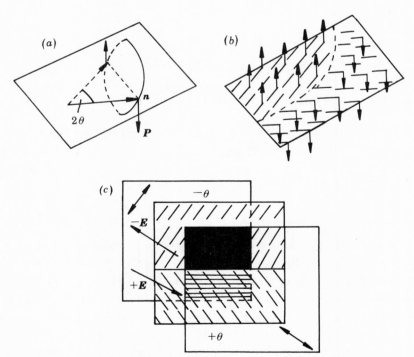

FIGURE 8. Operation of ferroelectric chiral smectic C* electro-optic devices: (a) two possible orientations on a surface imposing no specific direction; (b) dipole moments of the two possible orientations; (c) device operation with two crossed polarizers.

[104]

where I_0 is the parallel polarizer transmission. The resulting electro-optic effect has several very attractive features (Clark & Lagerwall 1980):

high speed of switching even at low voltages – for example 5 µs at 10 V;
bistability – both states are stable with long-term memory;
threshold behaviour – the switching shows a pronounced threshold.

The ferroelectric chiral smectic C effect has many potential applications, from fast shutters to complex multiplexed displays. However, the technological problems that need to be solved before this effect can be used are quite severe. The liquid crystal layers must be much thinner than those currently fabricated (2 µm rather than the usual 10 µm), and the surface alignment required is not readily accessible. Finally, no suitable chiral smectic C material exists with a room-temperature phase, a large transverse dipole moment and a tilt angle close to the preferred value ($\theta = 22.5°$).

($2\theta = 45°$?)

6. THERMO-OPTIC EFFECTS

I shall now consider the thermo-optic effects in liquid crystals that occur when the temperature of the liquid crystal is varied in the vicinity of a phase transition.

(a) Cholesteric materials

Another optical property of helically ordered liquid crystals is used in the thermo-optic (or thermochromic) effect in cholesteric materials (Elser & Ennulat 1976). In a well aligned planar sample, Bragg-like reflexions of one sense of circularly polarized light occur from the helical planes at a wavelength λ related at normal incidence to the average refractive index $\bar{n}$ and pitch P by

$$\lambda = \bar{n}P. \tag{11}$$

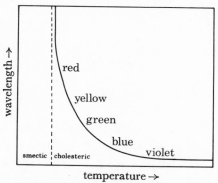

FIGURE 9. Temperature-dependence of selective reflexion from a cholesteric close to a smectic–cholesteric transition.

A strong thermochromic effect occurs just above the first-order phase transition from cholesteric to smectic phase, where P (and hence λ) diverges rapidly (figure 9). Thermochromic effects can be very sensitive, showing in some cases a colour shift through the visible spectrum for a change in temperature of only 0.5 K. Unlike all the other effects described in this paper, thermochromic devices are fabricated by printing an 'ink' containing the liquid crystal onto a substrate. The ink is produced by using one of two techniques. Droplets of the liquid crystal can be microencapsulated in polymer shells (diameters 5–50 µm) and the ink formed from a slurry of these capsules in water. Secondly, a dispersion of the liquid crystal droplets in a binder polymer film

[105]

can be used directly without the need to form capsules. There is a wide range of temperature sensing applications of thermochromic devices, for example thermometry, medical thermography, non-destructive testing, radiation sensing, fashion and advertising.

(b) Smectic A materials

Smectic A liquid crystals can adopt two metastable textures in a thin layer. For simplicity, if we consider a layer with homeotropic boundary conditions, there is a clear transparent texture with the director normal to the plates, and secondly there is a scattering texture shown schematically in figure 10. If the material has $\Delta\epsilon > 0$, it can be changed from one texture to the other by using the overlying nematic and isotropic phases with the application of heat and an electric field, as shown in figure 10. Heating to the isotropic phase followed by rapid cooling through the

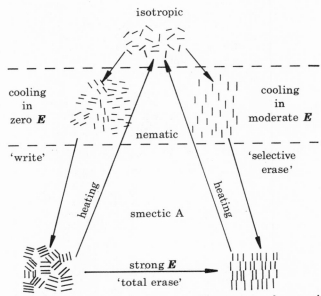

isotropic

cooling
in
zero **E**

cooling
in
moderate **E**

nematic

'write'

'selective
erase'

heating

heating

smectic A

strong **E**
'total erase'

FIGURE 10. Transitions between the clear and scattering textures of a smectic A layer with the use of an overlying isotropic phase.

nematic phase produces the scattering texture. Applying a moderate voltage as well as the heat produces alignment in the nematic phase and results in the aligned smectic texture. The application of a higher voltage erases the whole display by transferring the whole layer to the transparent texture. There are two quite distinct methods of applying the heat pulse.

(i) Laser addressing

A high-resolution image can be produced by scanning a focused laser beam across a smectic layer by using galvanometer mirrors and converting the laser energy into heat within the layer (Kahn 1973a) with the use of either absorptive surface layers (Kahn 1973a; Dewey 1980) or a dye dissolved in the liquid crystal (Hughes 1983). The system is shown schematically in figure 11, together with the projection optics for displaying the written image. The normal contrast is black written lines on a bright background, but this is reversed by the insertion of a Schlieren stop. Selective erasure is achieved by simultaneously scanning the laser and applying a voltage across the layer, and total erasure by the application of a higher voltage alone. The performance of laser-addressed systems is impressive with a possible resolution of up to several thousand lines in each axis. Writing speed depends upon the power of the laser used, and times as low as 3 s have been reported for writing over 10^6 picture elements with a high-power argon laser. There is almost indefinite storage, and capabilities exist for selective erasure, grey scale and multicolour display.

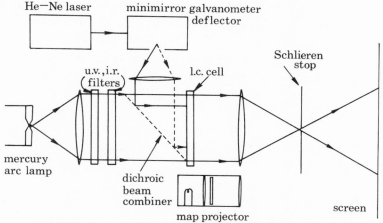

FIGURE 11. Construction of a transmissive laser-addressed smectic projection display.

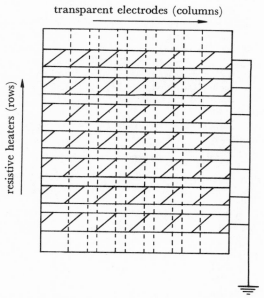

FIGURE 12. Electrode configuration of an electrically addressed smectic display.

(ii) *Electrical addressing*

The laser can be replaced as a source of heat by using an array of electrical heating elements (Hareng & Le Berre 1978). The heating elements (figure 12) are scanned sequentially and appropriate voltages applied to write or to erase the required information. The scattering can be observed directly (Hareng & Le Berre 1978) or a pleochroic dye can be dissolved to give an absorption contrast (Lu *et al.* 1982). Large displays, showing up to 100 alphanumeric characters by direct viewing, have been demonstrated (Lu *et al.* 1982; Le Berre *et al.* 1982) operating from

0 to 40 °C. The whole display is written in 2–5 s with a power consumption of 25 W, although the permanent memory significantly reduces the power consumption if only part of the information is changed.

7. CONCLUSIONS

There is a wide variety of electro-optic and thermo-optic effects in the nematic, cholesteric and smectic phases of thermotropic liquid crystals. Despite this wide variety there are a number of common features that form the building blocks of the various effects. In this paper many of these effects have been described in outline only; later papers, however, will examine displays based on the twisted nematic and pleochroic dye effects in significantly more depth.

REFERENCES

Assouline, G., Hareng, M. & Leiba, E. 1971 *Electron. Lett.* **7**, 699–700.
Berreman, D. W. 1972 *Phys. Rev. Lett.* **28**, 1683–1686.
Clark, M. G., Harrison, K. J. & Raynes, E. P. 1980 *Physics Technol.* **11**, 232–240.
Clark, N. & Lagerwall, S. 1980 *Appl. Phys. Lett.* **36**, 899–901.
Coates, D., Crossland, W. A., Morrissy, J. H. & Needham, B. 1978 *J. Phys.* D **11**, 2025–2034.
Crossland, W. A. & Ayliffe, P. J. 1982 *Proc. Soc. Inform. Display* **23**, 9–13.
de Gennes, P. G. 1968 *Solid State Commun.* **6**, 163–165.
de Jeu, W. H. 1980 *Physical properties of liquid crystal materials.* London, New York and Paris: Gordon & Breach.
Dewey, A. G. 1980 In *The physics and chemistry of liquid crystal devices* (ed. G. J. Sprokel), pp. 219–239. New York and London: Plenum Press.
Elser, W. & Ennulat, R. D. 1976 In *Advances in liquid crystals* (ed. G. H. Brown), vol. 2, pp. 73–172. New York, San Francisco and London: Academic Press.
Frank, F. C. 1958 *Discuss. Faraday Soc.* **25**, 19–28.
Freedericksz, V. & Zolina, V. 1933 *Trans. Faraday Soc.* **29**, 919–930.
Hareng, M. & Le Berre, S. 1978 In *Proc. IEDM, Washington,* Dec. 1978.
Heilmeier, G. H. & Zanoni, L. A. 1968 *Appl. Phys. Lett.* **13**, 91–92.
Heilmeier, G. H., Zanoni, L. A. & Barton, L. A. 1968 *Appl. Phys. Lett.* **13**, 46–47.
Helfrich, W. 1969 *J. chem. Phys.* **51**, 4092–4105.
Hughes, A. J. 1983 *RSRE Newsl. Res. Rev.* (In preparation.)
Janning, J. L. 1972 *Appl. Phys. Lett.* **21**, 173–174.
Kahn, F. J. 1973*a* *Appl. Phys. Lett.* **22**, 111–113.
Kahn, F. J. 1973*b* *Appl. Phys. Lett.* **22**, 386–388.
Le Berre, S., Hareng, M., Hehlen, R. & Perbet, J. N. 1982 *Soc. Inform. Display Digest* **8**, 252–253.
Lu, S., Davies, D. H., Albert, R., Chung, D., Hochbaum, A. & Chung, C. 1982 *Soc. Inform. Display Digest* **8**, 238–239.
Maier, W. & Meier, G. 1961 *Z. Naturf.* **16a**, 1200–1205.
Matsumoto, S., Kawamoto, M. & Kaneko, N. 1975 *Appl. Phys. Lett.* **27**, 268–270.
Mauguin, C. 1911 *Bull. Soc. Fr. Minér.* **34**, 71–117.
Meyer, R. B. 1968 *Appl. Phys. Lett.* **12**, 281–282.
Meyer, R. B. 1977 *Molec. Cryst. liq. Cryst.* **40**, 33–48.
Raynes, E. P. 1975 *Revue Phys. appl.* **10**, 117–120.
Raynes, E. P. 1979 *IEEE Trans. electron. Devices* **ED-26**, 1116–1122.
Raynes, E. P. & Shanks, I. A. 1974 *Electron. Lett.* **10**, 114–115.
Schadt, M. & Helfrich, W. 1971 *Appl. Phys. Lett.* **18**, 127–128.
Schiekel, M. F. & Fahrenschon, K. 1971 *Appl. Phys. Lett.* **19**, 391–393.
Steers, M. & Mircea-Roussel, A. 1976 *J. Phys., Paris* (Colloq. no. 3), **37**, 145–148.
Suzuki, K., Ishibashi, T., Satoh, M. & Isogai, T. 1981 *IEEE Trans. electron. Devices* **ED-28**, 719–723.
Tani, C. 1971 *Appl. Phys. Lett.* **19**, 241–242.
White, D. L. & Taylor, G. N. 1974 *J. appl. Phys.* **45**, 4718–4723.
Wysocki, J. J., Adams, J. & Haas, W. 1968 *Phys. Rev. Lett.* **20**, 1024–1025.

Phil. Trans. R. Soc. Lond. A **309**, 179–188 (1983)

Printed in Great Britain

Practical limits on addressing twisted nematic displays

By Barbara Needham

Standard Telecommunication Laboratories Limited, London Road,
Harlow, Essex CM17 9NA, U.K.

Reflective twisted nematic liquid crystal displays have gained very wide user acceptance in consumer and professional applications. They are compact flat-panel displays with a very low power consumption, which modulate ambient light and as a result have very good legibility in a wide range of illumination conditions. There is a need for such low-power flat-panel displays with high information content as data displays for portable equipment.

A number of problems must be overcome if twisted nematic displays are to be made complex enough for these applications. The most important factor determining the number of display elements that can be addressed is the viability of matrix-addressing techniques. This paper explores the limits in the number of lines that can be addressed for a display operating in reflexion.

THE BASIC OPERATION OF THE TWISTED NEMATIC DISPLAY

The basic structure and operation of the twisted nematic display is shown in figure 1. A thin layer of liquid crystal material (6–15 µm thick) of positive dielectric anisotropy is confined between two glass panels that have a transparent, electrically conductive layer on their interior surfaces overlaid by a surface alignment layer (e.g. rubbed polyimide), which imposes a preferred alignment direction on the liquid crystal director. The director at opposite surfaces is constrained to lie orthogonal in the plane of the surfaces, imposing a 90° twist within the liquid crystal layer. Incident polarized light has its electric vector waveguided by the liquid crystal on passing through the display, so that it lines up with the polarization direction of the lower polarizer and is transmitted (figure 1a). On applying a field somewhat greater than the threshold voltage to the liquid crystal, the director reorients as shown in figure 1b, lining up with the field. The wave-guiding of the twisted structure is therefore removed and polarized light is blocked by the lower polarizer, giving rise to the well known dark-on-light display.

The best display performance is achieved under direct drive, that is when each display element is individually driven. The complexity of direct-drive displays is limited by two main factors. One is the difficulty and expense of making connections to large numbers of picture elements, and the second is the cost of drive electronics, which would become prohibitive.

Addressing a display on a matrix format reduces the number of external connections and drivers from $n+1$ to a minimum of $2\sqrt{n}$ for a display of n picture elements. The motivation for matrix addressing is therefore clear.

MATRIX ADDRESSING

I shall consider a matrix addressed display in which one panel carries a number of parallel conductor stripes, the rows, and the other carries a second set orthogonal to the first, the columns. Suitable voltages applied to the rows and the columns cause voltages to appear across the liquid crystal at the intersection of the row and column electrodes, resulting in some optical change.

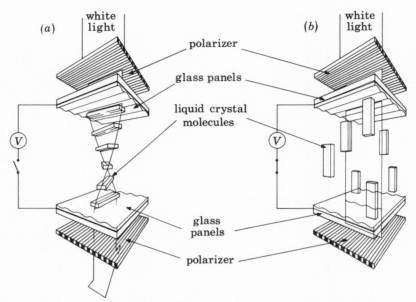

FIGURE 1. The twisted nematic field effect: (*a*) without field; (*b*) with field.

Static matrix addressing

The simplest way to address such a matrix is to apply various steady voltages to its row and column electrodes. The voltage across any matrix element is defined by the voltage difference between its row and column electrodes. Row and column voltages can be chosen to maximize the voltage difference between selected and unselected elements, but it is not possible to devise a scheme in which unselected elements have no voltage across them. Thus the display must have a nonlinear electro-optic response (i.e. a sharp threshold) so that unselected elements are not activated.

There are restrictions on the patterns that can be displayed by using such a static addressing scheme. For example a closed figure with a hollow centre cannot be displayed, since elements in the centre are energized at the same time as those around the perimeter. However, static addressing can sometimes be used to advantage in applications where the information displayed is restricted, such as for cross-hair graticules in optical instruments, or waveform identity addressing as devised by Shanks (1979) for oscilloscope displays.

Dynamic matrix addressing

The solution to the problem of addressing closed figures is to enter information to the display sequentially a line at a time. Figure 2 shows a schematic of such a time-multiplexed matrix addressing scheme. The information to be displayed is applied in parallel to the columns and an enable or scanning voltage (V_s) is applied to line 1. Information on the columns is then changed to that appropriate to line 2, and line 2 is enabled. The information is entered as the data voltage (V_d) such that $-V_d$ is applied to selected elements and $+V_d$ to unselected elements. Under these circumstances a selected element receives a pulse of magnitude $|V_s + V_d|$ while it is being addressed and pulses of magnitude $|V_d|$ while all the other lines are being addressed. Similarly an unselected element receives a pulse of magnitude $|V_s - V_d|$ while being addressed and $|V_d|$ for the remainder of the frame time. The drive voltage across any element therefore changes with time, so the transient

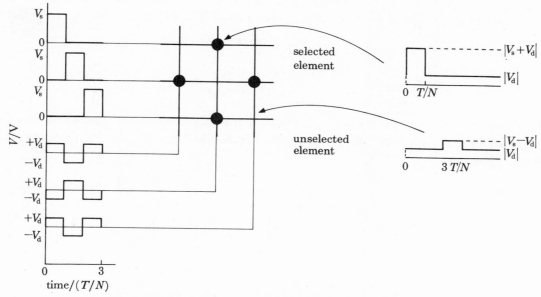

FIGURE 2. Schematic of a time-multiplexed addressing scheme
(three lines in a matrix of N lines shown, frame time $= T$).

response characteristics of the display are important in analysing matrix performance. For the twisted nematic display, rise and fall times are similar and fairly long (tens to hundreds of milliseconds). Since the frame time for the entire display must be faster than about 25 ms to avoid perceptible flicker, the steady-state response of the display is determined by the integral over the full-frame time of the drive waveform (r.m.s. voltage). This is known as a fast-scan multiplexed addressing scheme.

The optimum voltage levels for fast-scan multiplexing

The derivation of the optimum voltage levels V_s and V_d and calculation of the resultant voltage discrimination between selected and unselected elements for a particular display matrix requires consideration of the waveform across the elements during the entire scan. This has been analysed by Alt & Pleshko (1974). By taking the general case of a matrix of N lines the r.m.s. voltages across selected (V_{ON}) and unselected elements (V_{OFF}) can be calculated:

$$V_{ON}^2 = (V_s + V_d)^2/N + V_d^2 - V_d^2/N \qquad (1)$$

and
$$V_{OFF}^2 = (V_s - V_d)^2/N + V_d^2 - V_d^2/N. \qquad (2)$$

As the number of lines N increases, V_{ON} and V_{OFF} become dominated by the data voltage V_d. This is an inevitable consequence of fast-scan addressing. Thus for any particular device the number of lines that can be scanned depends on the sharpness of the electro-optic response curve.

[111]

A schematic electro-optic response curve relating optical response to the applied voltage, known as the optical transfer characteristic (o.t.c.), is shown in figure 3. A figure of merit or voltage discrimination ratio, M, between the voltage level $\hat{V}_{ON}$ when the display is judged to be in its minimum acceptable ON condition and the voltage $\hat{V}_{OFF}$ when the display is acceptably OFF can be defined as

$$M = \hat{V}_{ON}/\hat{V}_{OFF}. \tag{3}$$

From this ratio for a given o.t.c. with a figure of merit M, the maximum number of scanned lines N_{max} and the corresponding data and scanning voltage V_d and V_s can be completely specified:

$$N_{max} = \{(M^2+1)/(M^2-1)\}^2; \tag{4}$$

$$V_d = \tfrac{1}{2}V_{OFF}[M^2+1]^{\frac{1}{2}}; \tag{5}$$

$$V_s = \tfrac{1}{2}V_{OFF}(M^2+1)^{\frac{3}{2}}/(M^2-1). \tag{6}$$

For a given number of scanned lines N, the optimum choice in the ratio of scanning and data voltages results in maximum r.m.s. voltage discrimination between selected and unselected elements given by

$$\hat{V}_{ON}/\hat{V}_{OFF} = \{(N^{\frac{1}{2}}+1)/(N^{\frac{1}{2}}-1)\}^{\frac{1}{2}}. \tag{7}$$

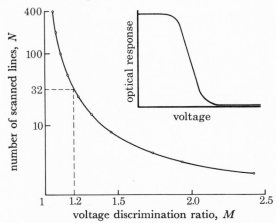

FIGURE 3. The voltage discrimination ratio, M ($= \hat{V}_{ON}/\hat{V}_{OFF}$) (defined from the optical transfer characteristic (o.t.c.)) required for an increasing number of scanned lines, N. The inset shows a schematic o.t.c.

This shows that increasing the number of scanned lines places increasingly stringent demands on the voltage discrimination ratio M of the display device (see figure 3). A large scanning capability N implies a low value for M. For successful operation with 32 lines, for example, the display must switch through its defined minimum acceptable contrast with only 20 % change in r.m.s. voltage. Thus the measurement of display contrast as a function of applied r.m.s. voltage (o.t.c.) shows the maximum number of lines N that can be addressed satisfactorily for a defined display contrast.

The application of the above results to a determination of the maximum number of lines that can be addressed for a real display is not straightforward. The sharpness of the o.t.c. can be optimized for twisted nematic displays through the liquid crystal material used and through a number of aspects of display construction. In addition the voltage discrimination ratio V_{ON}/V_{OFF} may be chosen in several different ways depending on the application.

The threshold voltage of the display is viewing-angle dependent, and light travelling obliquely through the display and bisecting the two surface alignment directions will show a decreased threshold voltage compared with normal incidence along the direction of tilt of the liquid crystal molecules. Figure 4 shows the o.t.c. measured in transmitted light for normal, 10° and 45° incidence in this direction, which is known as the principal viewing plane. The viewing angle required for a particular display application, together with the minimum acceptable contrast for selected elements and the maximum acceptable contrast for unselected elements, governs the choice of V_{ON} and V_{OFF}.

The threshold voltage is also temperature-dependent, with typical shifts of 5–10 mV/°C. The temperature range specified for a particular application is therefore an added restriction in multiplexability, unless electronic temperature compensation of the threshold voltage is acceptable.

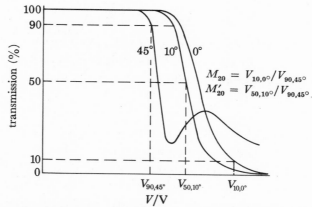

FIGURE 4. The optical transfer characteristic of a twisted nematic liquid crystal display measured in transmission at 0°, 10° and 45° incidence in the low threshold voltage quadrant.

The effect of display construction on multiplexability

A number of aspects of display construction influence the sharpness of the optical transfer characteristic. Their effects are widely accepted and have been summarized previously (Kahn & Birecki 1980; Birecki & Kahn 1980).

They may be summarized as a series of guidelines that should be followed to optimize contrast and viewing angle for maximum multiplexability.

(*a*) No barrier layer: a barrier layer acts as a capacitor in series with the liquid crystal layer, reducing the sharpness of the o.t.c.

(*b*) Thin low-tilt alignment layers (about 2° tilt angle): the sharpness of the threshold is reduced as the alignment tilt angle increases. However, a finite tilt angle is required to prevent reverse-tilt patches on switching.

(*c*) Minimum possible cell thickness (depends on liquid crystal birefringence – the Mauguin limit).

(*d*) Less than 90° twist angle (85°).

(*e*) No cholesteric additives: addition of a cholesteric additive alters the director configuration and optical properties of the layer.

(*f*) Alignment of the polarizer dielectric axes orthogonal to the adjacent surface alignment directions.

Optimization of liquid crystal material

A number of parameters of liquid crystal materials influence the sharpness of the o.t.c. and therefore the multiplexability. To achieve high levels of multiplexing, liquid crystals are required that have a low birefringence Δn, low splay:bend elastic constant ratio, k_{33}/k_{11}, low twist elastic constant (k_{22}) and long pitch (i.e. no cholesteric additive).

MATRIX ADDRESSING LIMITS FOR REFLECTIVE DISPLAYS

A severe limitation is imposed on the figure of merit, M, owing to shifts in the threshold voltage over the usable temperature range. Electronic temperature compensation for the threshold voltage is a requirement for achieving high levels of multiplexing. The definition of figure of merit, M, then depends only on the viewing angle and the contrast required. Conventionally this has been taken from the o.t.c.s measured in transmission.

The o.t.c. in reflexion

In practice the derivation of the o.t.c. for a reflective display is complicated by the superposition of contrast from the first and second passes through the cell. Light follows different paths through the cells on these passes, depending on the reflector and illumination conditions.

Reflective measurement system

A reflective optical bench has been built that was designed to give diffuse illumination conditions similar to those provided by diffused overhead ceiling lights, daylight, etc., in typical viewing conditions. The display cell mount is rotatable through 360° and is positioned at the focus of a hemispherical white diffusing reflector. Illumination is provided by a ring of filament lamps mounted on the base plane of the hemisphere but shielded to prevent direct illumination of the display. Front-surface reflexions of the diffusing reflector are eliminated by a black absorbing band around the viewing direction. This models the role of the head in typical viewing conditions. The display is imaged on to a detector by using a telescope and fibre-optic probe rotatable through $\pm 45°$ from normal incidence to the display.

A comparison was made between the figures of merit measured from the o.t.c.s of one display in transmission and in reflexion with a number of different reflector materials. The display cell measured was 8.5 µm thick, with a rubbed polymer surface alignment layer, and was filled with E60A (B.D.H. Chemicals Ltd). HN42 polarizers were laminated onto the display with their dichroic axes orthogonal to the adjacent surface alignment directions.

The normal incidence transfer characteristic is shifted to lower voltages when measured in reflected light, as is that measured at 10° incidence, and the oblique incidence threshold at 45° in the low-voltage quadrant is increased slightly. As a consequence the multiplexing figures of merit for 10:1 contrast at normal incidence, $V_{10,0°}/V_{90,45°}$, and for 2:1 contrast at 10° from normal incidence, $V_{50,10°}/V_{90,45°}$ (defined for less than 1.1:1 contrast for unselected elements at 45° incidence, as shown in figure 4), are lower by approximately 15%, increasing the apparent multiplexing capability. The increase in threshold voltage at oblique angles of incidence is because some of the light reflected in this direction had its first pass in the high threshold voltage quadrant. The reduction in threshold voltage observed at 10° and normal incidence is also observed in the high-voltage quadrant and the two adjacent quadrants. This is due to the shadow cast on the reflector by a first pass through the display in the low-voltage direction. Thus diffuse reflected illumination has the effect of considerably widening the viewing angle of a multiplexed display in addition to causing an apparent increase in threshold sharpness. Both specular and fairly diffuse reflectors show very similar results.

Berman & Chan (1980) investigated the role of the shadow cast on the reflector on the viewability of twisted nematic displays under direct drive, and came to similar conclusions.

These factors are important in assessing the ultimate limits of matrix addressing of practical reflective twisted nematic displays.

Assessment of a multiplexed twisted nematic display

A 14-way multiplexed display was chosen for this purpose. This is a 160-character display of 2.8 mm character height (four lines of 40 characters on a 7×5 dot matrix) developed at S.T.L. Two rows of characters (14 lines) are addressed from the top of the display and two rows from the bottom. It was designed for use in home or office environments for a viewing angle between 0° and 45° with electronic temperature compensation.

Aspects of cell construction were optimized to take full advantage of developments in multiplexable liquid crystal materials. A polyimide alignment layer was chosen because its conformal properties give advantages in forming a barrier in very thin layers (25–40 nm). This was rubbed unidirectionally to give a low tilt angle (2°) alignment in directions biased to give an 85° twist angle in the assembled display. The display was sealed to produce a cell spacing of about 10 μm and filled with a selected liquid crystal material. Polarizers were laminated onto the front and rear surfaces with their dichroic axes aligned orthogonal to the adjacent surface alignment directions.

The choice of liquid crystal material was dictated by the overriding requirement for a sharp threshold. It has been found (Bradshaw & Raynes 1982) that the addition of alkyl esters to biphenyls decreases k_{33}/k_{11}, particularly for long-chain alkyl esters (Sturgeon, this symposium). The main limitation for such materials is the temperature of formation of smectic phases. The dielectric anisotropy is also decreased, leading to a higher threshold voltage. This may lead to an increased threshold sharpness, but generally the effects are expected to be small (Kahn & Birecki 1980). The higher threshold and resultant operating voltage were acceptable for our application.

Two liquid crystal multiplexing materials that are mixtures of long-chain alkyl esters and biphenyls were developed in cooperation with B.D.H. Chemicals. The data for these materials for 11 μm thick cells are shown in table 1. The first of these materials, LC14, was developed for the 14-way multiplexed display, which requires a voltage discrimination ratio ($V_{ON}/V_{OFF} = M$) of 1.315. As discussed previously, the choice of V_{ON} and V_{OFF} depends on the requirements of the user. The specified viewing angle for the display was 0° to 45°. Two figures of merit are taken from the optical transfer characteristics measured in transmission. M_{20} (20 °C) is defined for 10:1 contrast at normal incidence for selected elements and less than 1.1:1 contrast for unselected elements at 45°. M'_{20} is defined for a reduced contrast of 2:1 at 10° from normal incidence for selected elements. M_{20} was 1.68 and M'_{20} was 1.31. In reflexion these are reduced to about 1.39 and 1.16 respectively, owing to shadowing effects.

A number of these reflective displays were assessed under 14-way multiplexed drive in diffuse ambient illumination provided by overhead fluorescent lamps and daylight. Almost full display contrast was achieved from normal incidence out into the principal viewing plane for selected elements (i.e. $V_{ON} \approx V_{10,0°}$). Unselected elements were judged to have unacceptably high contrast for angles greater than 45° in the principal viewing plane (i.e. $V_{OFF} \approx V_{90,45°}$), but in spite of crosstalk the display was readable out to 55° incidence. Owing to the influence of the shadow the display was easily readable with better than 2:1 contrast out to glancing incidence in all directions outside the principal viewing plane.

TABLE 1. MEASUREMENTS ON TWO LIQUID CRYSTAL MIXTURES
IN CELLS 11 µm THICK AT 20 °C

	LC14	LC21
smectic–nematic transition temperature/°C	−20	−10
nematic–isotropic transition temperature/°C	60	60
birefringence (Δn)	ca. 0.146	ca. 0.146
V_{10}, normal incidence	3.46	3.34
V_{50}, 10° incidence	2.70	2.69
V_{90}, 45° incidence	2.06	2.11
$M_{20} = V_{10}/V_{90}$	1.68	1.58
$M'_{20} = V_{50}/V_{90}$	1.31	1.27
$-dV_{90}/dt$ (0–40 °C)/(mV/°C)	20	25

Relating this performance to the measured figures of merit in reflexion for the liquid crystal materials used, we would expect M_{20} to describe the threshold sharpness. M_{20} is 1.39, implying a maximum number of multiplexable lines, N_{max}, of 10. Experience with viewer trials shows that this underestimates N_{max} for a practical reflective display, illustrating the difficulty of judging perceived display characteristics on the basis of simple physical measurements. The alternative figure of merit, M'_{20}, measured in transmission (1.31 for LC14) predicts 14-way multiplexing for a 2:1 contrast from 10° incidence out into the principal viewing plane; this is clearly an underestimate of contrast and viewing angle for a practical reflective display. In practice M'_{20} is a convenient figure of merit to use because the underestimation of N_{max} due to transmissive measurements and the overestimation due to lower contrast and viewing angle compensate, so that it predicts N_{max} for good contrast between 0° and 45° for a reflective display.

The variation of threshold voltage with temperature for LC14 is larger than for most commercially available liquid crystals, but is fairly linear. As a result, electronic temperature compensation with a thermistor was found to be adequate.

LC21 was a mixture developed later with a larger proportion of the long-chain alkyl esters. This has an improved threshold sharpness at the expense of a shorter nematic temperature range and a larger variation of threshold voltage with temperature. An M'_{20} of 1.27 implies a maximum number of scanned lines of 18; 21-way multiplexing for a 6-line alphanumeric display would require an M'_{20} of 1.25. However, it is judged that with LC21 a 21-way multiplexed reflective display can be made with good contrast over a slightly narrower viewing angle than our 14-way multiplexed display and with contrast greater than 2:1 to glancing incidence outside the principal viewing plane. Because of the larger nonlinear change of threshold voltage with temperature, the thermistor sensor may be inadequate for electronic temperature compensation, and more sophisticated methods such as those proposed by Hilsum et al. (1978) may be necessary.

Limits in matrix addressing of twisted nematic displays

Kahn & Birecki (1980) derived performance limits for optimized fast-scan multiplexed twisted nematic displays by means of computer calculations of the optical transfer characteristics. They investigated the effect on their figures of merit of altering one parameter at a time, neglecting the effects of temperature, and derived any improvement in multiplexing that could be achieved from optimizing them. Their baseline was a 3° tilt angle, 90° twisted cell 8 µm thick, filled with E7 and doped with cholesteric to prevent reverse twist. The liquid crystal parameters varied were k_{22}, k_{33}/k_{11}, $\Delta\epsilon$ and Δn. In addition they varied cell thickness and twist angle, and investigated the effect of cholesteric dopants and polarizer orientation.

They estimated the upper limit for multiplexing for a theoretical material with all parameters optimized by adding improvements for each onto the baseline value for multiplexability. For the range of parameters investigated this requires $\Delta n = 0.1$, $k_{33}/k_{11} = 0.2$, $k_{22} = 2 \times 10^{-12}$ N and a pitch of infinity (no cholesteric dopant). The upper limit was $N = 24$ for temperature-compensated direct-view displays with more than 2:1 contrast between normal incidence and 40° in the principal viewing plane. These conditions were determined on the basis of measurements in transmitted light, but they state that viewing angles should be better in reflexion.

TABLE 2. MATERIAL PARAMETERS AND MULTIPLEXABILITY OF
A NUMBER OF LIQUID CRYSTAL MIXTURES

	LC14	LC21	E120	E130	E140
$T_{\mathrm{SN}}/°\mathrm{C}$	-20	-10	< -20	< -20	< -20
$T_{\mathrm{NI}}/°\mathrm{C}$	60	60	60	64.3	63
Δn (589.6 nm, 20 °C)	0.146	0.146	0.146	0.154	0.147
k_{33}/k_{11}†	ca. 0.95	ca. 0.9	1.04	0.88	0.90
M'_{20}	1.31‡	1.27‡	1.29§	1.27§	1.22§
N_{max}	14	18	16	18	25

† Reduced temperature of 0.9.
‡ Measured at 11 µm (S.T.L.).
§ Measured at 7 µm (B.D.H.).

We have found that display contrast, viewing angle and apparent threshold sharpness are considerably improved in reflexion. It is therefore useful to relate the results obtained for practical displays in reflexion to these calculated limits of performance. A number of material parameters of high-level multiplexing liquid crystals, including LC14, LC21 and more recent developments from B.D.H. (E120–E140), are listed in table 2 with their multiplexing figures of merit. The elastic constant ratios k_{33}/k_{11} of E120–E140 were kindly provided by M. J. Bradshaw at R.S.R.E. The effect of reducing k_{33}/k_{11} and the birefringence Δn on increasing the threshold sharpness is apparent. Relating these parameters to Kahn & Birecki's results and substituting for k_{33}/k_{11} and Δn, a maximum of approximately 10 lines would be expected for LC14 from their figure of merit for a contrast greater than 2:1 between 0° and 45°. In practice the 14-way multiplexed reflective display containing LC14 has high contrast within this angle and more than 2:1 contrast to glancing incidence.

Assuming that liquid crystal materials may be developed in the future with parameters approaching the Kahn & Birecki optimum, up to 14 more lines may be multiplexed than can be achieved with materials such as LC14. Displays multiplexed 32 ways have been demonstrated with acceptable viewing characteristics for many applications. The viewing cone of the displays is reduced to about 30°. On this basis 46 lines or more may ultimately be addressable, and 32 lines are addressable with existing materials.

Materials such as LC21 and E140 (B.D.H.), which have been developed more recently, have figures of merit M'_{20} that imply that 18 lines can be multiplexed in cells 10–11 μm thick and 25 lines in cells 7 μm thick with the same contrast and viewing angles as the 14-way multiplexed display described. These show that progress is being made towards ultimately achieving 46 lines or more of multiplexing for twisted nematic displays.

CONCLUSIONS

It has been shown in this paper that reflective-mode twisted nematic displays operating with good contrast over a 30° viewing cone may ultimately be multiplexed over about 46 lines. This is based on the assumption that liquid crystal materials will be developed with optimum characteristics ($k_{33}/k_{11} = 0.2$, $k_{22} = 2 \times 10^{12}$ N, $\Delta n = 0.1$) for multiplexing. The number is derived from the results of Kahn & Birecki (1980), taking account of the difference in characteristics of displays measured in transmission and reflexion and related to viewer trials on a 14-way multiplexed display made at S.T.L. For a 7×5 dot matrix format alphanumeric display, at least 12 lines of information could be displayed by dividing the matrix and addressing half the display from each side. For applications where a more restricted viewing angle is acceptable, this may be increased further. Assuming up to 400 columns for a display the maximum information content expected from fast-scan multiplexed twisted nematic displays is likely to be of the order of 5×10^4 picture elements.

I thank S.T.L. for permission to publish this paper. The work has been carried out with the support of the Procurement Executive, U.K. Ministry of Defence, sponsored by D.C.V.D.

REFERENCES

Alt, P. M. & Pleshko, P. 1974 *IEEE Trans. Electron. Devices* **ED-21**, 246–155.
Berman, A. L. & Chan, S. O. 1980 In *The physics and chemistry of liquid crystal devices*, pp. 241–252. New York and London: Plenum Press.
Birecki, H. & Kahn, F. J. 1980 In *The physics and chemistry of liquid crystal devices*, pp. 125–142. New York and London: Plenum Press.
Bradshaw, M. J. & Raynes, E. P. 1982 *Molec. Cryst. liq. Cryst.* (In the press.)
Hilsum, C., Holden, R. J. & Raynes, E. P. 1978 *Electron. Lett.* **14**, 430–432.
Kahn, F. J. & Birecki, H. 1980 In *The physics and chemistry of liquid crystal devices*, pp. 79–93. New York and London: Plenum Press.
Shanks, I. A. & Holland, P. A. 1979 *S.I.D. Int. Symp. Digest of Technical Papers*, pp. 112–113.

Phil. Trans. R. Soc. Lond. A **309**, 189–201 (1983)
Printed in Great Britain

Guest–host devices using anisotropic dyes

By T. J. Scheffer

Brown Boveri Research Center, CH–5405 Baden/Dättwil, Switzerland

An up-to-date review is given of Heilmeier and White–Taylor guest–host displays with their many variations, i.e. positive and negative dielectric anisotropy, positive and negative dichroism, flat and perpendicular boundary orientation, single and double cells, etc. Performance characteristics such as brightness, contrast ratio, operating voltage, field of view and multiplexability are compared with those of state-of-the-art twisted nematic devices. Applications for guest–host displays are discussed and new guest–host materials are suggested to further improve the performance of guest–host displays.

1. Introduction

Guest–host displays operate by the absorption of light by dichroic dye molecules oriented in a liquid crystal. The brightness of the display is varied by applying an electric field, thus changing the orientation of the liquid crystal. The original guest–host displays had to be operated at high temperatures, but they served to demonstrate the feasibility of the guest–host effect (Heilmeier & Zanoni 1968; Heilmeier *et al.* 1969). It was not until after the invention of the twisted nematic display (Schadt & Helfrich 1971) that suitable room-temperature host materials were developed. Since then many other guest–host schemes have been investigated. Probably the most important has been the scheme without polarizers described by White & Taylor (1974). In 1977 photo-chemically stable anthraquinone dyes with high dichroic ratios became available (Pellatt *et al.* 1980), and today ready-to-use black guest–host mixtures are commercially available.

It has been several years since these last advances were made, and yet guest–host displays have not progressed very far beyond the prototype stage. To understand why guest–host displays are not being manufactured in large quantities, we should investigate the performance of present prototype guest–host displays and compare it with that of commercially available twisted nematic displays. For most applications the positive qualities of guest–host displays, which include wide viewing angle, fewer polarizers and high brightness, are more than compensated for by the positive qualities of the twisted nematic display that must be sacrificed: high contrast, multiplexability and low-voltage operation. Twisted nematic displays are still to be preferred to guest–host displays for most applications. In this paper I shall review the state-of-the-art performance of guest–host displays in terms of viewing angle, contrast ratio and multiplexability. This approach will help to define the applications to which guest–host displays would be best suited. Another goal of this paper will be to extrapolate from present trends in the development of materials and technology to the future performance of guest–host displays.

A large number of guest–host schemes have been described in the literature. I shall not attempt to review all these schemes here, but rather shall cover only the more important effects involving nematic and cholesteric hosts. Smectic guest–host schemes (Pelzl *et al.* 1979; Lu *et al.* 1982) will not be discussed.

2. Heilmeier display

A non-twisted Heilmeier-type guest–host display using present cell technology is illustrated in figure 1. A rubbed polymer layer (not shown) gives flat boundary orientation with approximately 1° of pretilt and glass fibre spacers (also not shown) distributed over the display area maintain a uniform layer thickness, e.g. 8.0 ± 0.2 μm. A single polarizer is arranged in front of the layer so that its transmission direction is parallel to the rubbing direction on the adjacent cell plate. The cell is filled with a liquid crystal mixture having positive dichroism and positive dielectric anisotropy. In the absence of an applied voltage, the E-vector of normally incident light is nearly parallel to the optical axis, the direction of maximum absorption, and the cell appears dark. Where an electric field is present in the cell (the region of electrode overlap in figure 1), the optical axis in the layer is distorted and, in the middle of the layer, approaches a

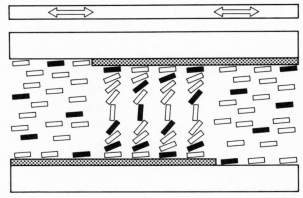

FIGURE 1. Sketch of original Heilmeier guest–host display. Transmission direction of polarizer is indicated by arrows.

direction parallel to the field. Here the E-vector of the incident light is nearly parallel to the direction of minimum absorption and the field-on region of the cell appears bright. This type of guest–host display is referred to as a negative contrast display because it shows light–coloured symbols on a dark background.

The transmission–voltage curve and the angular dependence of the contrast ratio are important parameters of the display performance, which can be computed. The orientation of the local optical axis in the layer can be determined from the elastic continuum theory (Deuling 1978; Scheffer 1981). The optical transmission of the distorted layer can then be determined from Maxwell's equations by approximating the layer as a stack of birefringent, dichroic slices of constant thickness, having uniform orientation of the optical axis within each slice (Berreman 1973).

Figure 2 compares the transmission characteristic computed for twisted (discussed in §3) and non-twisted Heilmeier displays (solid line and dotted line) with that of a twisted nematic display (broken line). The cell parameters and material constants used for this computation are summarized in table 1. For simplicity the absorption constants and refractive indices are assumed to be independent of wavelength. The parallel and perpendicular polarizer transmission constants listed in table 1 refer to the transmission of linearly polarized light by a single polarizer (Scheffer & Nehring 1977). The reduced-voltage scale on the x axis represents the applied voltage normalized by the corresponding Freedericksz threshold voltage. The guest–host curves have been normalized to make the upper border the asymptote of the transmission for infinite reduced voltage.

It is seen in figure 2 that the slope of the transmission characteristic for the non-twisted Heilmeier display is much more gradual than that of the twisted nematic display. A good measure of the steepness of the transmission curve, and hence the suitability for multiplexing, is the ratio of the voltage at 50 % of the optical response to the Freedericksz threshold voltage. For the twisted

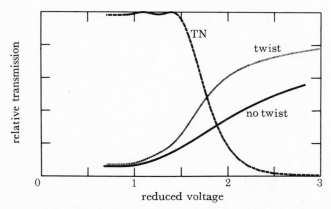

FIGURE 2. Transmission characteristic computed for twisted and non-twisted Heilmeier display compared with that for twisted nematic display. Upper border is transmission asymptote for guest–host curves.

TABLE 1. MATERIAL CONSTANTS† AND CELL PARAMETERS
USED IN COMPUTATION

splay elastic constant	k_{11}	11.98×10^{-12} N
twist elastic constant	k_{22}	6.48×10^{-12} N
bend elastic constant	k_{33}	17.11×10^{-12} N
parallel dielectric constant	$\epsilon_{\parallel}$	15.49
perpendicular dielectric constant	$\epsilon_{\perp}$	4.35
extraordinary refractive index	n_{e}	1.640
ordinary refractive index	n_{o}	1.492
extraordinary absorption constant	α_{e}	0.3940 $\mu\mathrm{m}^{-1}$
ordinary absorption constant	α_{o}	0.0394 $\mu\mathrm{m}^{-1}$
Freedericksz k_{11} threshold voltage	U_{o}	1.095 V
parallel polarizer transmittance	$T_{\parallel}$	0.7360
perpendicular polarizer transmittance	$T_{\perp}$	0.0016
layer thickness	d	8 μm
pretilt at boundaries	α_{0}	1°

† The material constants k_{11}, k_{22}, k_{33}, $\epsilon_{\parallel}$ and $\epsilon_{\perp}$ were measured at 20 °C by H. Schad (personal communication 1982) on the host material ZLI-1840, from Merck, Germany.

nematic cell of figure 2 this ratio is 1.78, while for the non-twisted Heilmeier cell it is 2.69. The response is less steep for the Heilmeier display because the transmission depends on the orientation of the optical axis throughout the entire layer, and the regions near the cell boundary continue to absorb light until comparatively high voltages are applied. The transmission of the twisted nematic cell depends mainly on the orientation of the optical axis in the central region of the layer, which is nearly vertical for an applied voltage of 2–3 times the threshold voltage. The gradual slope of the guest–host curve means that the Heilmeier-type display is not suitable for multiplex drive even for 2:1 ratios, and that higher voltages are required for direct drive. Materials possessing a lower k_{33}/k_{11} ratio or a smaller $(\epsilon_{\parallel} - \epsilon_{\perp})/\epsilon_{\perp}$ will not significantly increase the steepness of the transmission curve. Schadt (1982) has shown that it is possible to multiplex the Heilmeier guest–host display, at least theoretically, up to a factor of 30:1 by employing a dual-frequency addressing scheme.

Figure 3 compares the iso-contrast diagrams computed for the twisted nematic and non-twisted Heilmeier displays. These diagrams give the contour lines of the constant contrast ratio of 10:1 (good) and 4:1 (lower limit) on a polar coordinate system. These iso-contrast diagrams are analogous to the conoscopic figures seen under the microscope. The azimuthal viewing angle extends completely around the diagram from 0 to 360° and the angle of incidence, measured in air, corresponds to the radial direction starting in the centre of the diagram for normal incidence and extending out to the periphery for grazing incidence at 90°. It is clear from figure 3 that the field of view for acceptable contrast ratio is considerably wider for the Heilmeier guest–host display than for the twisted nematic display.

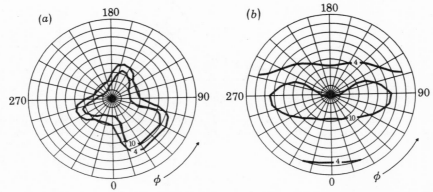

FIGURE 3. Iso-contrast diagrams computed for (a) twisted nematic and (b) Heilmeier guest–host displays in transmission. Case shown is for reduced voltage = 3.0 V.

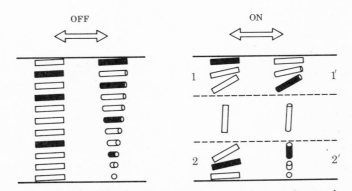

FIGURE 4. Sketch showing optical axis orientation in boundary regions of twisted and non-twisted Heilmeier cells.

3. HEILMEIER DISPLAY WITH 90° TWIST

Uchida *et al.* (1979*a*) have demonstrated that the transmission characteristic of the Heilmeier display can be made steeper by incorporating a 90° twist in the layer. This can be explained by referring to figure 4. In the field-off state (left side) the twisted layer absorbs nearly the same amount of light as the non-twisted layer because, within the Mauguin limit, the elliptical polarized modes propagating through the layer depart only very slightly from linear polarized modes

(Nehring 1982). The major axis of the ellipse follows very nearly in step with the twisted structure and therefore remains more or less parallel to the optical (absorbing) axis. In the field-on state (right side), however, the twisted layer absorbs significantly less light than the non-twisted layer. The reason for this is that the orientation of the optical axis in the boundary region 2' on the lower substrate of the twisted cell is at right angles to the E-vector of the polarized light and is therefore non-absorbing.

The transmission curve computed for the twisted Heilmeier display (dotted line in figure 2) is much steeper at intermediate voltages than for the non-twisted display (solid line), but the approach to saturation remains gradual at higher reduced voltages because the upper boundary region 1' still absorbs light. The 50 % ratio for this case is 1.86, indicating that very limited multiplexing is possible. We computed the iso-contrast diagram for this case to see if the 90° twist had a detrimental effect, but, aside from skewing the contour lines to some extent, the field of view remained about the same.

4. GUEST–HOST DISPLAY WITH NEGATIVE DICHROIC MIXTURES

In the previous sections I have dealt with guest–host mixtures with positive dichroism, i.e. where the extraordinary absorption constant α_e is larger than the ordinary absorption constant α_o. Pelzl *et al.* (1979); Demus *et al.* (1979); Schadt (1979) and Uchida & Wada (1981) have investigated guest–host displays that employ systems with negative dichroism, i.e. $\alpha_e < \alpha_o$. These workers employed derivatives of tetrazine dyes that had been elongated by 3,6 substitution. The direction of the optical transition dipole moment of these dyes is perpendicular to the plane containing the tetrazine ring (and perpendicular to the long molecular axis).

Negative dichroic mixtures can be used to produce positive contrast displays, i.e. displays in which dark symbols are shown against a light background. Very steep transmission curves have been measured for such displays (Pelzl *et al.* 1979). To compute the transmission characteristic for such a negative dichroic system we assumed the parameters given in table 1, except for the absorption constants, which we assumed as $\alpha_e = 0.068$ and $\alpha_o = 0.394$, independent of wavelength. We were therefore considering a black mixture exhibiting a dichroic ratio $R' = \alpha_o/\alpha_e$ $= 5.8$, which is the same as the dichroic ratio we obtained for the red tetrazine dye mixture Lixon GR-63w available from Chisso, Japan. As before, a single polarizer is oriented in front of the layer with its transmission axis parallel to the rubbing direction of the adjacent cell plate. The computed transmission characteristics in figure 5 are indeed much steeper than those of positive dichroic systems (figure 2). This seems reasonable because the change in absorption for a small change in tilt angle is much greater when the direction of maximum absorption is pointing toward the observer rather than when it is at right angles to the observer. The cell with the 90° twist has the steepest curve (dotted line) giving a voltage ratio of 1.22 at 50 % optical response. The transmission characteristic for the non-twisted layer (solid line) has a voltage ratio of 1.29, which is still much lower than the value of 1.78 found for twisted nematic displays (broken line). The twisted layer has a steeper response because, referring to figure 4, the twisted configuration of the lower cell plate allows the lower boundary region 2' to absorb light in the field-on state, whereas the boundary region 2 is always non-absorbing. The iso-contrast diagram computed for this case (not shown) indicates that an acceptable contrast ratio can be observed through a wide range of viewing angles.

These displays would have probably replaced twisted nematic displays in many applications if a black, negative dichroic mixture with a dichroic ratio of 10 had been available. For reasons of dichroic ratio and colour, however, it seems very doubtful that such a hypothetical mixture will ever be achieved.

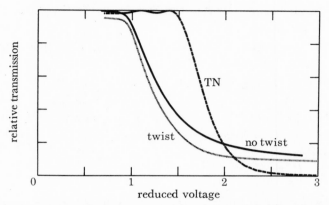

FIGURE 5. Transmission characteristic computed for twisted and non-twisted Heilmeier display with negative dichroism, compared with that for twisted nematic display.

5. GUEST–HOST DISPLAY WITH NEGATIVE DIELECTRIC ANISOTROPY

A positive contrast display can also be achieved in a positive dichroic mixture possessing a negative dielectric anisotropy. In this case the optical axis orients itself perpendicularly to the electric field, which requires a homeotropic orientation of the optic axis in the layer. A small pretilt from the layer normal ensures a defined tilt direction along the transmission axis of the polarizer when the field is applied. This pretilt can be introduced by oblique evaporation of SiO followed by treatment with a homeotropic surfactant (Koshida 1981; Uchida *et al.* 1979*b*), or by other methods such as treatment with a homeotropic surfactant followed by unidirectional rubbing (Uchida *et al.* 1980). Uchida & Wada (1981) have found pretilt angles of 3–5° to be most suitable.

We computed the transmission characteristic for this display by using the material constants measured for the nematic mixture Lixon EN-24 from Chisso, Japan (Schad 1982). We assumed extraordinary and ordinary absorption constants of 0.483 and 0.0483 and a pretilt angle of 1° away from the layer normal. Layer thickness and polarizer characteristics are given in table 1. The transmission characteristic shown in figure 6 is quite steep; the voltage ratio for 50 % optical response is 1.23, which is comparable with the transmission characteristics computed for the negative dichroic mixtures. This is reasonable, because just above threshold the axis of maximum absorption is perpendicular to the layer, where it can produce a large change in absorption for a small change in tilt angle. Saturation is only gradually approached because higher voltages are required before the boundary layers begin to absorb light. Introducing a 90° twist in the layer does not improve the transmission characteristic for this case.

The field of view for this device is illustrated by the iso-contrast diagrams of figure 7 computed for both transmission and reflexion (computed with half the dye concentration). In the transmissive mode an acceptable contrast ratio can be obtained over a wide range of viewing angles, but in the reflective mode the wide range of viewing angles is spoiled by 'holes' where the contrast is reversed. The birefringence of the layer is responsible for this effect. The holes occur at the angle of incidence for which the homeotropic layer behaves like a quarter-wave plate. Polarized light incident at this angle is completely extinguished at the azimuthal angles of 45°, 135°, 225° and 345° where the extraordinary and ordinary rays have equal amplitudes. Assuming an 8 μm layer thickness and a birefringence of 0.083 we computed this angle of incidence to be 42.0° (Françon 1956). This effect is also present in the turned-on state of the Heilmeier-type displays

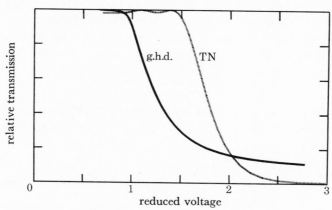

FIGURE 6. Transmission characteristic computed for guest–host display with negative dielectric anisotropy compared with that for twisted nematic display.

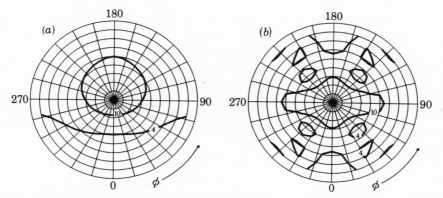

FIGURE 7. Iso-contrast diagrams computed for (a) transmissive and (b) reflective guest–host display with negative dielectric anisotropy. Case shown is for reduced voltage = 3.0 V.

(§§ 2 and 3), but it is not as apparent because it only affects the character segments rather than the whole display background. These holes do not appear with negative dichroic dyes (§ 4) because the homeotropic state is the dark state and in the parallel-oriented state only one of the optical modes is excited.

New mixtures are required to take advantage of the excellent multiplexibility of this effect. Stable nematic hosts as well as dyes with adequate solubilities and dichroic ratios in these new hosts will have to be synthesized. The host materials must be free of ionic impurities to prevent the occurrence of electrohydro-dynamic effects in the display. Furthermore, a new cell technology based on pretilted homeotropic alignment would have to be developed.

6. REFLECTIVE GUEST–HOST DISPLAYS WITH QUARTER-WAVE PLATE

Cole & Kashnow (1977) investigated a reflective guest–host scheme that employs a metallic reflector behind a quarter-wave retardation plate oriented with its optical axis at 45° with respect to the optical axis of a parallel-oriented guest–host layer. In the field-off state the guest–host cell acts like a polarizer, and this combination strongly absorbs light. In the field-on state the polarization effect is turned off and the layer appears bright. This display is brighter than the other guest–host displays discussed so far because no polarizing sheet is present. However, the contrast ratio is lower because the dichroic ratio of the guest–host layer (*ca*. 10) is not as high as the dichroic ratio present in commercial polarizers (10–65) (Scheffer & Nehring 1977).

Uchida & Wada (1981) suggest that the field of view of this display will be restricted because of the angular dependence of the quarter-wave plate. However, the iso-contrast diagram that we computed for this display indicates that an acceptable contrast ratio can be observed for all angles of incidence up to 50°, regardless of the azimuthal angle. We found that the wavelength variation of the phase retardation is a minor effect when the quarter-wave plate is matched to the peak wavelength response of the eye occurring at 550 nm. This display effect is not only limited to the parallel Heilmeier guest–host cell. A positive contrast device, for example, can be realized by using the display cells described in §§4 or 5.

7. DOUBLE-LAYERED GUEST–HOST DISPLAYS

A technologically more complex guest–host scheme that also does not require a polarizer is the double-layer scheme (Uchida *et al.* 1981; Sawada & Masuda 1982). Two parallel-oriented Heilmeier guest–host cells are placed over each other so that the directions of the optical axes on adjacent faces are at right angles. In the field-off state the system acts like a pair of crossed polarizers and no light is transmitted; application of an electric field turns off the absorption. This display has identical electrode patterns in each cell, which are activated in pairs. To minimize parallax, the display is constructed with three glass plates, the central one being in contact with the liquid crystal on both sides. The twisted layers of §3 would also work with this scheme and improve the steepness of the transmission characteristic. A positive contrast display could be made by using the guest–host layers described in §§4 and 5.

Double-layered guest–host displays absorb unpolarized light more efficiently than any other guest–host system without polarizers. Nevertheless, the contrast ratio is not as high as for single-polarizer guest–host schemes because of the low dichroic ratio of the guest–host layer. Double-layered guest–host displays operating in the reflective mode can employ either a metallic reflector or a diffuse, depolarizing reflector.

8. WHITE–TAYLOR DISPLAY

In the display devices discussed so far, the optical modes propagating through the guest–host layer are linearly or almost linearly polarized. In order to absorb unpolarized light efficiently, these displays require an additional external element, such as a polarizer, quarter-wave plate or an additional guest–host layer. The White & Taylor (1974) guest–host display differs in that it does not require any additional element. In this display the liquid crystal has a highly twisted structure in the field-off state, obtained by doping the nematic with a chiral additive. The optical modes propagating in a layer with short pitch are highly elliptical and therefore very efficiently absorb unpolarized light. More details about the absorption and reflexion of light from twisted structures can be found in the article by Nehring (1982).

White–Taylor guest–host displays can be made with either homeotropic or flat boundary orientation. The type of boundary orientation has a large effect on the performance of the display.

(a) Cells with homeotropic orientation

The field-off state of a cell with homeotropic orientation adopts the 'scroll' texture (Kawache *et al.* 1978), which very slightly scatters light. An advantage of homeotropic orientation is that the scroll texture is re-established without the formation of disclinations when the voltage across the layer is switched off, even if there is a relatively large number of turns of the cholesteric helix within the layer. Because it is possible to have a large number of turns in the layer, the optical absorption is more efficient (shorter pitch and more circular modes) than with flat orientation, even though the boundary regions in the homeotropic cell make almost no contribution to the absorption.

At relatively low applied voltages this texture transforms to the highly scattering focal-conic texture, where the helical axis is oriented in the plane of the layer. At higher voltages the pitch of the helix increases until above a certain critical voltage the pitch becomes infinite and a homeotropic structure is adopted within the whole layer. This state only weakly absorbs light. The homeotropic state becomes metastable in a certain range of voltages below this critical voltage. A number of ingenious schemes have been developed that make use of this hysteresis effect to obtain high multiplex rates in phase-change devices without added dye (Ohtsuka *et al.* 1973; Tsukamoto & Ohtsuka 1974; Alder & Shanks 1976; Walter & Tauer 1978). All these schemes employ a holding voltage within the bistable range where either of two optical states can be present. Within the bistable range the homeotropic state slowly reverts to the focal–conic state through a nucleation process initiated by dust particles and other disturbances in the cell. This means the display must be continually refreshed, which limits the number of lines that can be multiplexed. This refreshing and the rather low contrast that exists between the focal conic and homeotropic states when dye is added have prevented their practical application for guest–host displays.

(b) Cells with flat orientation

The planar Grandjean texture is adopted in layers with flat or nearly flat boundary orientation. Practical devices must have less than about two turns of the cholesteric helix because layers that contain more turns exhibit disturbing after-images when the display element is switched off, caused by light scattering from disclinations. With only one or two turns in the layer, the operating voltage can be low, but the contrast ratio is not as high as that achieved with homeotropic boundaries because of the longer pitch. The layer tolerance must be rigidly controlled in these displays to avoid the occurrence of Grandjean–Cano disclinations, which are highly visible because of the stepwise change in brightness appearing across them.

Hysteresis behaviour is also observed with flat boundary orientation, but here the transmission characteristic exhibits a true threshold voltage, above which the grid pattern distortion occurs (de Zwart *et al.* 1979). At higher voltages this pattern gives way to the focal-conic fingerprint texture, and at still higher voltages a uniform untwisted structure is established. The approach to saturation is gradual because of the boundary regions, which reorient only at relatively high voltages. A very important advantage of this display is that it can be multiplexed to a limited

[127]

extent by using conventional waveforms (Suzuki *et al.* 1981). Aftergut *et al.* (1981) have measured the angular dependence of the contrast ratio in layers with many helical turns and find good agreement with Saupe's (1980) computations. No azimuthal dependence was observed. We computed a significant amount of azimuthal dependence in layers with fewer helical turns, in agreement with de Zwart's (1982) measurements.

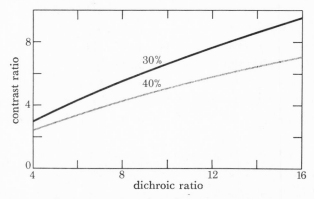

FIGURE 8. Computed dependence of contrast ratio on the dichroic ratio for single-turn White–Taylor display. Brightness of field-on state is held constant at 30 and 40 %. Voltage applied is five times the Freedericksz k_{11} threshold.

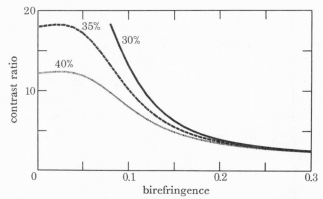

FIGURE 9. Computed dependence of contrast ratio on host birefringence for single-turn White–Taylor display. Brightness of field-on state is held constant at 30, 35 and 40 %. Voltage applied is five times the Freedericksz k_{11} threshold.

To predict the extent to which improvements in materials will influence the performance of this type of guest–host display, we computed the dependence of contrast ratio on the dichroic ratio of the black mixture for the constant on-state brightnesses of 30 % and 40 %. For this computation we used the material constants $k_{33}/k_{22} = 2.0$, $k_{33}/k_{11} = 1.5$, $(\epsilon_{\parallel} - \epsilon_{\perp})/\epsilon_{\perp} = 2.5$, $n_{e} = 1.64$ and $n_{0} = 1.49$ for an 8.0 μm thick cell with one helical turn and a metallic reflector. A voltage corresponding to five times the k_{11} Freedericksz threshold was applied to the layer. Figure 8 shows that the efforts required to increase the dichroic ratio from its present value of 10 to 12 or to even 14 are likely to be rewarded with only a marginal increase in the contrast ratio of the display. This insensitivity to the dichroic ratio results from the fact that neither the field-on nor the field-off brightness is determined by a single absorption constant but always by both α_{e} and α_{o}. Under these conditions even an infinite dichroic ratio would give a finite contrast ratio. It should be emphasized that the conclusions drawn from figure 8 apply only to the multiplexable guest–host display with flat boundary orientation described above. Figure 8 does not apply to cells with homeotropic boundary orientation or to cells with flat boundary orientation operating at very high applied voltages.

The computed dependence of the contrast ratio on the host birefringence (figure 9) for a dichroic ratio of 10, i.e. present values, shows that large gains in contrast ratio can be obtained by decreasing the birefringence of the host material. The black mixture ZLI-1841 (Merck, Germany), has a birefringence of 0.15. According to figure 9 contrast ratios could be approximately doubled while maintaining the same brightness by employing a host material with a birefringence of 0.10. The contrast ratios that can be obtained with very low birefringence would correspond to the contrast ratios that can be obtained with the double cells discussed in §7.

9. Cholesteric display with negative host

Analogous to the Heilmeier display, a positive-contrast White–Taylor–type display can be made by employing a host material with a negative dielectric anisotropy and a cell with homeotropic boundaries (Gharadjedaghi 1981; Gharadjedaghi & Voumard 1982). To maintain a homeotropic orientation in the field-off state, the thickness:pitch ratio d/p must satisfy the requirement $d/p < k_{33}/2k_{22}$, which limits the number of turns of the helix in the field-on state to about one. For a practical display, allowing for thickness and temperature variations, this means that there will be less than one turn of the cholesteric helix in the layer, and so a host material with a low birefringence is absolutely necessary. No hysteresis is observed in the transmission characteristic and the curves are steep enough to allow for limited multiplexing (Nagae et al. 1981). The cell technology is simpler than with the scheme discussed in §4 because no pretilt from homeotropic orientation is required.

10. Quasi-positive schemes

Quasi-positive guest–host schemes refer to those guest–host effects that inherently give negative contrast but that can be made to appear like a positive contrast display. Gharadjedaghi & Saurer (1980) and Uchida & Wada (1981) have examined several quasi-positive schemes in which the boundary alignment is different in the regions where the displayed symbols appear (either on one or both substrates) than on the remaining background regions. Gharadjedaghi & Saurer (1980) also describe a quasi-positive White–Taylor guest–host scheme where a few micrometres of glass has been etched out from one of the substrates under the symbols to make the guest–host layer thicker there. Quasi-positive behaviour can also be obtained by inversely driving the display, i.e. addressing the background region and those elements that are not to be seen and applying zero voltage to those regions that are to appear dark (Scheffer & Nehring 1980). A particularly simple quasi-positive scheme employs the White–Taylor effect with homeotropic boundaries in a cell with two compartments (Oh & Kramer 1982).

11. Applications

Guest–host displays are finding application in areas where they perform better than twisted nematic displays or where it is impossible to employ twisted nematic displays. An example of the former case is the liquid crystal oscilloscope display (Shanks et al. 1979), which has recently become a commercial product (Scopex Instruments Ltd, private communication 1982). Here the full viewing angle of a White–Taylor display with scattering reflector is used with a 128×256 point matrix because no multiplexing is required for single-valued oscilloscope waveforms. The guest–host effect is also finding application in reflective-mode matrices where the liquid crystal is in direct contact with thin-film transistors (Ymasaki *et al.* 1982; Crossland *et al.* 1982) and varistors (Levinson *et al.* 1982), where an opaque substrate prohibits placement of the rear polarizer required for the twisted nematic effect. These devices are still at the prototype stage, but the market potential is large. Guest–host displays also have some potential application for car dashboards where a wide viewing angle is essential and polarizers are not suitable because of their poor environmental stability under conditions of high humidity and high temperature. However, these devices are slower than twisted nematic devices because of the increase in the host viscosity due to the dissolved dye. This increase in viscosity can slow down the display response to such an extent that the display no longer satisfies present safety regulations, which specify a response time of about 1 s at $-30\,^{\circ}\mathrm{C}$. However, the requirements for car clocks and car radios are not so stringent and application of guest–host displays in these areas is possible today. Even more applications are expected when improved guest–host mixtures become available.

References

Aftergut, S. & Cole H. S. 1981 *Appl. Phys. Lett.* **38**, 599–601.
Alder, C. J. & Shanks, I. A. 1976 *Electron. Lett.* **12**, 326–327.
Berreman, D. W. 1973 *J. opt. Soc. Am.* **63**, 1374–1380.
Cole, H. S. & Kashnow, R. A. 1977 *Appl. Phys. Lett.* **30**, 619–621.
Crossland, W. A., Ayliffe, P. J. & Ross, P. W. 1982 *Proc. Soc. Inform. Display* **23**, 15–22.
Demus, D., Kruecke, B., Kuschel, F., Nothnick, H. U., Pelzl, G. & Zaschke, H. 1979 *Molec. Cryst.* **56**, 115–121.
Deuling, H. J. 1978 In *Solid state physics* (ed. L. Liebert), vol. 14 (*Liquid crystals*), pp. 77–107. New York: Academic Press.
de Zwart, M. 1982 *Proc. Electronic Display Conference* 1982, London, 5–7 October.
de Zwart, M. & van Doorn, C. Z. 1979 *J. Phys. Paris* **40** (suppl. C3), 278–284.
Françon, M. 1956 In *Handbuch der Physik* (ed. S. Fluegge), vol. 24, pp. 171–460. Berlin: Springer-Verlag.
Gharadjedaghi, F. 1981 *Molec. Cryst. liq. Cryst.* **68**, 127–135.
Gharadjedaghi, F. & Saurer, E. 1980 *IEEE Trans. Electron. Devices* **ED–27**, 2063–2069.
Gharadjedaghi, F. & Voumard, R. 1982 *J. appl. Phys.* **53**, 7306–7313.
Heilmeier, G. H., Castellano, J. A. & Zanoni, L. A. 1969 *Molec. Cryst. liq. Cryst.* **8**, 293–304.
Heilmeier, G. H. & Zanoni, L. A. 1968 *Appl. Phys. Lett.* **13**, 91–92.
Kawachi, M., Kato, K. & Kogure, O. 1978 *Jap. J. appl. Phys.* **17**, 1245–1250.
Koshida, N. 1981 *J. appl. Phys.* **52**, 5534–5536.
Levinson, L., Castleberry, D. & Becker, C. 1982 *J. appl. Phys.* **53**, 3859–3864.
Lu, S., Davies, D. H., Albert, R., Chung, D., Hochbaum, A. & Chung, C. 1982 *Society for Information Display digest of technical papers*, vol. 13, pp. 238–239.
Nagae, Y., Isogai, Y., Kawakami, H. & Kaneko, E. 1981 In *Proc. 1st European Display Research Conference*, pp. 51–54.
Nehring, J. 1982 *J. chem. Phys.* **75**, 4326–4337.
Oh, C. S. & Kramer, G. 1982 *Proc. Soc. Inform. Display* **23**, 23–27.
Ohtsuka, T., Tsukamoto, M. & Tsuchiya, M. 1973 *Jap. J. appl. Phys.* **12**, 371–378.
Pellatt, M. G., Roe, I. H. C. & Constant, J. 1980 *Molec. Cryst. liq. Cryst.* **59**, 299–316.
Pelzl, G., Schubert, H., Zaschke, H. & Demus, D. 1979 *Kristall Technik* **14**, 817–823.
Saupe, A. 1980 *J. chem. Phys.* **72**, 5026–5031.
Sawada, K. & Masuda, Y. 1982 *Society for Information Display digest of technical papers*, vol. 13, pp. 176–177.
Schad, H. 1982 *Society for Information Display digest of technical papers*, vol. 13, pp. 244–245.
Schadt, M. 1979 *J. chem. Phys.* **71**, 2336–2344.
Schadt, M. 1982 *Appl. Phys. Lett.* **41**, 697–699.
Schadt, M. & Helfrich, W. 1971 *Appl. Phys. Lett.* **18**, 127–128.
Scheffer, T. J. 1981 In *Advances in liquid crystal research and applications* (ed. L. Bata), vol. 2, pp. 1145–1153. New York: Pergamon Press.

Scheffer, T. J. & Nehring, J. 1977 *IEEE Trans. Electron. Devices* **ED–24**, 816–822.

Scheffer, T. J. & Nehring, J. 1980 In *The physics and chemistry of liquid crystal devices* (ed. G. J. Sprokel), pp. 173–198. New York: Plenum.

Shanks, I. A., Holland, P. A. & Smith, C. J. T. 1979 *Displays* April, pp. 33–41.

Suzuki, K., Ishibashi, T., Satoh, M. & Isogai, T. 1981 *IEEE Trans. Electron. Devices* **ED–28**, 719–723.

Tsukamoto, M. & Ohtsuka, T. 1974 *Jap. J. appl. Phys.* **13**, 1665–1666.

Uchida, T., Ohgawara, M. & Wada, M. 1980 *Jap. J. appl. Phys.* **19**, 2127–2136.

Uchida, T., Seki, H., Shishido, C. & Wada, M. 1979a *Molec. Cryst. liq. Cryst.* **54**, 161–174.

Uchida, T., Seki, H., Shishido, C. & Wada, M. 1979b *IEEE Trans. Electron. Devices* **ED–26**, 1373–1374.

Uchida, T., Seki, H., Shishido, C. & Wada, M. 1981 *Proc. Soc. Inform. Display*, **22**, 41–46.

Uchida, T. & Wada, M. 1981 *Molec. Cryst. liq. Cryst.* **63**, 19–44.

Walter, K. & Tauer, M. 1978 *IEEE Trans. Electron. Devices* **ED–25**, 172–174.

White, D. L. & Taylor, G. N. 1974 *J. appl. Phys.* **45**, 4718–4723.

Ymasaki, T., Kawahara, Y., Sunichi, M., Kamamori, H. & Nakamura, J. 1982 *Society for Information Display digest of technical papers*, vol. 13, pp. 48–49.

Phil. Trans. R. Soc. Lond. A **309**, 203–216 (1983)

Printed in Great Britain

Numerical modelling of twisted nematic devices

By D. W. Berreman

Bell Laboratories, 600 *Mountain Avenue, Murray Hill, New Jersey* 07974, *U.S.A.*

Over most of each active region in nematic and chiral nematic twist cells the motion and configuration of the liquid crystal layer does not vary appreciably with position parallel to the surfaces. In such laminar regions the statics, dynamics and optics of the cell can be accurately simulated at low cost on a computer of moderate size, given the appropriate physical parameters. Methods and recent advances in simulation of laminar regions are reviewed. Bistable twist cells are simulated for illustration. Important problems of stability and edge effects in the presence of electric fields await solution with two- or three-dimensional simulations.

Introduction

In recent years many laboratories have set up computer programs to simulate the behaviour of nematic and chiral nematic displays. Several of these laboratories have published results of investigations into the effects of varying parameters on the static optical performance of twist cells (Baur 1980, 1981; van Doorn *et al.* 1980; Birecki & Kahn 1980). Simulation of dynamic behaviour has also been done and has led to an understanding of reverse-twist and optical bounce effects (van Doorn 1975; Berreman 1975 *a*). Interesting effects from 'weak anchoring' of directors on cell surfaces have been predicted by such computer simulation (Berreman 1980) but they remain largely unrealized (Yang 1983).

Twist cells with very high multiplexing numbers are used in some modern hand-held calculators. Computer simulation shows how parameters may be optimized to achieve such high multiplexing numbers (Baur 1980, 1981; van Doorn *et al.* 1980; Birecki & Kahn 1980).

We are currently interested in bistable chiral nematic twist cells (Heffner & Berreman 1982; Berreman & Heffner 1982). Successful computer simulation of their static, dynamic and optical behaviour has greatly enhanced our understanding of the operation of these cells. The first dynamic simulation of a bistable twist cell was done before one was made.

Many twist cells are made with chiral dopants in an ordinary nematic liquid crystal. Our experience suggests that the dopant concentration is usually low enough for the elastic, dielectric and optical parameters to be close to those of the nematic host. However, the usual methods of measuring these parameters are not applicable when the material is chiral. A comparison of static measurements of the optical behaviour or capacitance in twist cells with computer predictions would provide a means for measuring elastic, dielectric and optical parameters of chiral mixtures.

Modelling cells with one-dimensional variation of directors

Computer simulation of twist cell behaviour is done in two stages. First the orientation of the nematic director as a function of distance from each cell surface is computed with a particular set of parameters. For equilibrium states the required parameters are the mean-square electric displacement field, the two dielectric and three elastic constants and the natural helicity, if the nematic is chiral, and the orientation of the director at each surface. The electrical capacitance per unit area in the cell is a useful by-product of this computation of director orientation. The mean-square voltage is determined by the displacement and the capacitance. It is also easy to integrate the components of the energy density to obtain the energies per unit area in the cell.

Once the director configuration is found, the optical transmission and reflexion may be computed as a function of wavelength and viewing direction. This requires knowledge of the anisotropic optical parameters of the liquid crystal. Guest–host displays may easily be simulated by using complex numbers for the optical parameters to include absorption. Polarizers, conductive coatings, glass and even a reflective layer may be included in the sandwich for optical computations. So long as scattering is negligible, these computations may be done quickly on a computer by using a 4 × 4 matrix method (Smith 1965; Berreman & Scheffer 1970). Scattering must be treated separately.

To compute dynamic effects an initial configuration, which may come from an equilibrium computation, and five viscosity parameters are needed in addition to the previously listed parameters. There are very few liquid crystals for which the five viscosities have been determined. However, comparisons of observed dynamic variation of optical transmission or cell capacitance with predictions from fluid dynamic computations are possible and would provide a means for determining or verifying estimates of these viscosities.

There are two classes of numerical methods for finding director configurations in regions where variation parallel to the surfaces is negligible. One class is based on direct integration of the Euler–Lagrange differential equations for minimum Helmholtz free-energy per unit area in the liquid crystal. In this case the director orientation and its rate of change with z position is estimated at one point, such as one surface. The differential equations are then integrated to the opposite surface by a numerical method. If the integration does not give the desired orientation at the second surface, the initial conditions are altered, within the restrictions set by the problem, until it does.

The second is the class of relaxation methods. This class may include the full dynamic problem or it may be abbreviated by ignoring transverse motion of the fluid. In these computations an initial configuration is assigned to the directors and the configuration is then adjusted according to certain equations of motion that cause the energy in the liquid crystal to relax toward a minimum.

INTEGRATING STATIC EULER–LAGRANGE EQUATIONS

The strain free-energy density of a chiral nematic liquid crystal is described by the Oseen–Frank equation (Oseen 1933; Frank 1958; de Gennes 1975). When the director varies only in the z direction, normal to the suface, this free energy may be written as follows in Cartesian and polar coordinates:

$$F_s = \tfrac{1}{2}k_{11}(n_z')^2 + \tfrac{1}{2}k_{22}(n_x n_y' - n_y n_x' - 2\pi/P)^2 + \tfrac{1}{2}k_{33}\{(n_z n_x')^2 + (n_z n_y')^2 + (n_x n_x' + n_y n_y')^2\}$$
$$= \tfrac{1}{2}k_{11}(\theta'\sin\theta)^2 + \tfrac{1}{2}k_{22}(\beta'\sin^2\theta - 2\pi/P)^2 + \tfrac{1}{2}k_{33}\cos^2\theta\{\theta'^2 + (\beta'\sin\theta)^2\}.$$

k_{ii} are elastic coefficients, n_i are rectangular director components and P is the 360° pitch of the unstrained chiral nematic. The angle β is the azimuth and θ is the tilt measured from normal to the surfaces. Primes represent differentials with respect to z. The electrostatic free-energy density is

$$F_e = \frac{D^2}{2\epsilon_0\{\epsilon_\parallel n_z^2 + \epsilon_\perp(n_x^2 + n_y^2)\}}$$
$$= \frac{D^2}{2\epsilon_0(\epsilon_\parallel \cos^2\theta + \epsilon_\perp \sin^2\theta)},$$

[134]

where D is the displacement field and $\epsilon_\perp$ is the degenerate component of the dielectric tensor. The Helmholtz free-energy density is the strain free-energy plus the electrostatic free-energy: $F_H = F_s + F_e$. Equilibrium configurations are such that the integral of the Helmholtz free-energy density from one surface to the other has an extreme value, subject to the conditions imposed at the boundaries (see, for example, Thurston & Berreman 1981).

The Gibbs free-energy density is the strain free-energy density minus the electrostatic energy density. The integral, G, of the Gibbs free-energy with fixed applied voltage is also extremal (see, for example, Thurston & Berreman 1981). However, since it is necessary to know the configuration before the local voltage gradient can be defined, this fact is not useful in finding configurations.

The Euler–Lagrange equations to obtain extremal integrals of the Helmholtz free-energy are least complicated in polar coordinates. They are

$$\frac{\partial F_H}{\partial \theta} - \frac{d}{dz}\left(\frac{\partial F_H}{\partial \theta'}\right) = 0,$$

and, because F_H does not depend explicitly on β,

$$\partial F_H / \partial \beta' = T,$$

where T is a constant of integration and is also a measure of torque about the polar axis.

The preceding Euler–Lagrange equations may be manipulated to give

$$\theta'' = \left(\frac{\sin\theta\cos\theta}{k_{11}\sin^2\theta + k_{33}\cos^2\theta}\right)\left\{(k_{33} - k_{11})\,\theta'^2 + 2k_{22}\beta'(\beta'\sin^2\theta - 2\pi/P)\right.$$
$$\left. + k_{33}\beta'^2(\cos^2\theta - \sin^2\theta) + \frac{D^2(\epsilon_\parallel - \epsilon_\perp)}{\epsilon_0(\epsilon_\parallel\cos^2\theta + \epsilon_\perp\sin^2\theta)^2}\right\}$$

and

$$\beta' = \frac{T/\sin^2\theta + k_{22}\,2\pi/P}{k_{22}\sin^2\theta + k_{33}\cos^2\theta}.$$

These equations combine appropriate parts from the Leslie equations (Leslie 1970) for chiral nematics in magnetic fields and the Deuling equations (Deuling 1974) for twisted ordinary nematics in electric fields.

A disadvantage of polar coordinates is that special care must be taken if the director comes close to parallel with the polar axis, $\theta = 0$. The difficulty is with the second, β' equation, which gives a divergent value at $\theta = 0$ unless the torque, T, is zero. Fortunately θ will be exactly zero only at the surfaces, if at all, in non-trivial static problems. In that case it is known in advance that no torque can be maintained and the integration actually becomes simpler. When θ only comes close to zero we maintain accuracy in the integration by shortening the steps so that the change in β per step does not exceed a few degrees.

HUNTING FOR STATIC CONFIGURATIONS

Once a routine for integrating the equations for director configuration is written it is still necessary to provide an efficient means to estimate the initial conditions necessary to obtain a solution consistent with the desired boundary conditions. This will be illustrated by an example. Suppose we are considering a chiral nematic cell with directors attached parallel to the surface at one surface ($\theta_i = 90°$) and at an angle $\theta_f = 55°$ at the other. Suppose further that the director at the second surface is turned $\frac{3}{4}$ turn with respect to the first ($\beta_f = 270°$). Let the elastic constants and dielectric constants be those for E7, as listed in table 1. Let us find a configuration at 1.43 V.

[135]

(In our program we actually aim towards $V^2 = 2.045$.) Before we can start the integration we must estimate the rates of change of θ and β at the first surface and also the displacement field. (In practice we estimate θ', $\beta' \sin \theta$ and D^2 because the final results vary more smoothly with these functions when θ or D is small.)

TABLE 1. ASSUMED PARAMETERS FOR CHIRAL DOPED E7 NEMATIC MIXTURE AT 20 °C

splay, twist and bend elastic constants (Raynes *et al.* 1979)
  $k_{11} = 11.7 \times 10^{-12}$ N $k_{22} = 8.8 \times 10^{-12}$ N $k_{33} = 19.5 \times 10^{-12}$ N
extraordinary and ordinary dielectric constants (Raynes *et al.* 1979)
  $\epsilon_\parallel = 19.5$ $\epsilon_\perp = 5.17$
viscosity parameters (twice those for MBBA) (de Jeu 1978)
  $\gamma_1 = 0.190$ Pa s $\eta_1 = 0.242$ Pa s $\eta_2 = 0.0476$ Pa s $\eta_3 = 0.0832$ Pa s $\eta_{12} = 0.0130$ Pa s
optical parameters for 0.6328 μm light
  $n_o = 1.513$ $n_e = 1.728$

After a few guesses we can find four different tries that lead to approximately the desired configuration, as illustrated in figure 1. Then we assume that the final results are approximately linearly dependent on the initial estimates:

$$\theta_f \approx a_{11} + a_{12}D^2 + a_{13}\theta' + a_{14}\beta' \sin \theta,$$

$$\beta_f \approx a_{21} + a_{22}D^2 + a_{23}\theta' + a_{24}\beta' \sin \theta,$$

and
$$V^2 \approx a_{31} + a_{32}D^2 + a_{33}\theta' + a_{34}\beta' \sin \theta.$$

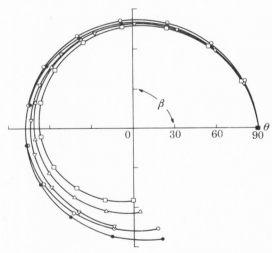

FIGURE 1. Search for a configuration at 1.43 V in an asymmetric cell with $\theta_i = 90°$, $\theta_f = 55°$, $\beta_f = 270°$, thickness: pitch ratio $h/P = 0.5$. Hunting is done by fixing θ_i and adjusting θ', $\beta' \sin \theta$ and D^2. Liquid crystal parameters are given in table 1. Marks are at equally spaced levels in the cell.

With the four tries we obtain four sets of parameters for these three approximate equations, enough to determine the 12 values of a_{ij}. We may then invert the equations and get a linear estimate of the initial conditions that would give the desired values of θ_f, β_f and V^2. The actual integration with the newly found initial parameters will miss the mark slightly but the process may be repeated, replacing a poor estimate with the newest one. Five or six iterations of this procedure, which is easily automated, usually gives the desired results of the integration to one part in 10 000.

Once one solution is found it is only necessary to use four of the previous approximate solutions and alter the target value of V^2 to find another solution nearby. This process can be continued to get the variation of solutions with V^2 without much further human intervention.

If plots of various functions against V are to be made, it is often unnecessary to seek solutions at a particular value of V. In those cases we select a series of values of D^2, use only three trial solutions and the first two of the three preceding equations with a_{i2} omitted and let the computed values of V fall where they may. This procedure avoids a problem that arises if there are multiple solutions at a single voltage, since the solutions are usually monotonic functions of D^2.

SPECIAL TREATMENT FOR SYMMETRICAL CELLS

In cells that have the same director tilt at each surface, configurations at moderate voltages are usually symmetrical about the mid-plane. At the mid-plane either $\theta = 90°$ or $\theta' = 0$, and β is half its final value. It is useful to start the integration at the centre of the cell to find such solutions because the trajectories are less sensitive to small errors in launching parameters there. If θ' is zero, we adjust the initial value of θ rather than θ' but otherwise the procedure is the same as if we started at one surface. Such a search is illustrated in figure 2, for a cell with $\beta_t = 360°$ and surface tilts of $55°$.

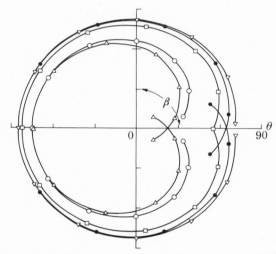

FIGURE 2. Search for configuration at 1.86 V in a symmetric twist cell with boundary tilts $\theta_i = \theta_t = 55°$, $\beta_t = 360°$, and $h/P = 1$. Hunting was done by using $\theta' = 0$ at the mid-plane and adjusting θ, $\beta' \sin \theta$ and D^2. Liquid crystal parameters are from table 1.

There is also one analytic solution to the trajectory equations in symmetrical cells that may be used as a starting configuration in place of one found by trial and error. This configuration is one of constant tilt θ and constant β'. Hence

$$\beta' = (\beta_t - \beta_i)/h$$

and the first Euler–Lagrange equation in polar coordinates may be solved for D^2, giving

$$D^2 = \left\{ \frac{\epsilon_0(\epsilon_\parallel \cos^2 \theta + \epsilon_\perp \sin^2 \theta)^2}{(\epsilon_\parallel - \epsilon_\perp)} \right\} \beta' \{ 2k_{22}(2\pi/P - \beta' \sin^2 \theta) + k_{33}\beta'(\sin^2 \theta - \cos^2 \theta) \}.$$

[137]

Even if D^2 computed from the preceding equation is negative so that D is imaginary, the solutions may still be continued into the physically meaningful region. The voltage on the cell is

$$V = \frac{hD}{\epsilon_0(\epsilon_\parallel \cos^2\theta + \epsilon_\perp \sin^2\theta)}.$$

USING STATIC CONFIGURATION RESULTS

To illustrate the usefulness of static configuration computations I shall carry through with more computations on the two cells considered in the preceding section.

It is often very instructive to make plots of Gibbs free-energy in the cell as a function of the square of the applied voltage (see figures 3 and 4). At zero voltage the plot is a line of nearly constant slope. The slope depends on the capacitance of the cell when in its field-free configuration. The curve then passes through a region of transition and finally follows another line of nearly constant slope that depends on the capacitance when the directors are mostly aligned nearly parallel to the electric field, if the liquid crystal has positive dielectric anisotropy, or perpendicular to it if the anisotropy is negative. In cells with high multiplexing numbers the change in slope is very abrupt. In bistable cells such curves double back to make a loop as illustrated in figures 3 and 4. The top of the loop is formed by unstable equilibrium states that are higher in energy than two adjacent bistable states at the same voltage, one of which is on the steeper and one on the gentler sloping line.

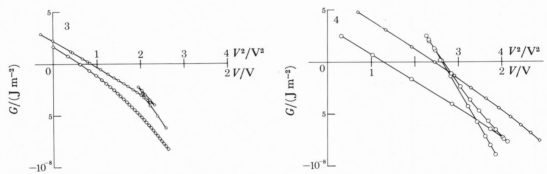

FIGURE 3. Gibbs free-energy plotted against V (nonlinear bottom scale) and V^2 (linear top scale) for the cell in figure 1 if it is 10 μm thick. Notice that the topologically separate states ($\diamond$) with 90° twist have lower energy than the bistable states (o) for this configuration. If they can be initiated, the bistable configurations eventually decay through disclination lines to this 90° twist state.

FIGURE 4. Gibbs free-energy plotted against V^2 for cell described in figure 2 if it is 10 μm thick. $\diamond$, A topologically separate set of states with final $\beta = 180°$, invariant mid-plane tilt 90° and final tilt 125°, which could be reached by the same cell through a disclination line if it were energetically preferred.

Because of the ambiguity in direction of a director, one should always investigate solutions that differ from the expected one by half a turn, and that have a final tilt that is the supplement of the expected value, as shown at the right in figure 7. Such alternative solutions are not accessible from the others through continuous director motion that is invariant parallel to the surfaces. However, they may form through the passage of a disclination line across the picture element. This will occur if the alternative solution has a lower energy than the intended ones, as in figure 3. If the undesired state is lower in energy only briefly during switching of the cell, it may not have sufficient time to develop. Nevertheless it is desirable to avoid such possibilities altogether since transitions through disclinations are usually slow to reverse.

Another curve that is instructive is a plot of the director tilt angle at the centre of the cell as a function of applied V^2. In the usual twist cell of positive dielectric anisotropy this angle (measured from normal to the surfaces) will approach zero as the voltage increases. In a cell with high

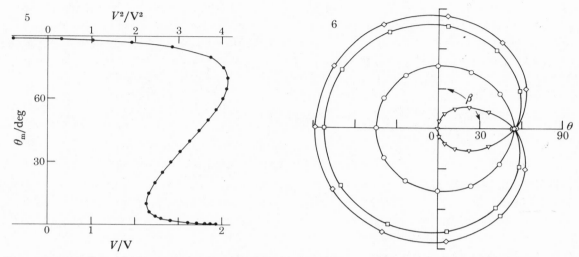

FIGURE 5. Mid-plane tilt θ_m plotted against V^2 (linear top scale) and V (bottom scale). Data from the four configurations of the preceding figure are included. The cell is bistable between about 1.5 and 2.1 V if nothing nucleates a transition to the state of lower energy through a moving 'wall'.

FIGURE 6. Three configurations at 1.86 V and one at no field ($\diamond$). The inner (first) and third loops are stable; the intermediate one is unstable. The first loop is the UP state; the third the DOWN state. Parameters are as in the previous figure. The outermost loop is the only stable solution at no field with 360° twist.

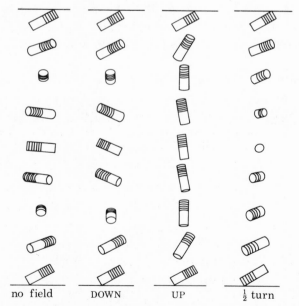

no field   DOWN   UP $\frac{1}{2}$ turn

FIGURE 7. Illustration of director configurations for the three stable states of figure 6 and an alternative half-turn state. DOWN designates the state at 1.86 V with directors most nearly parallel to the surfaces, and UP that with directors most nearly normal. Tilt near the centre in the UP state is exaggerated.

[139]

multiplexing numbers the angle will change very abruptly when some optimal operating voltage is reached. In a bistable cell this effect is exaggerated and the curve doubles back on itself, with the region of positive slope representing the unstable states between the stable ones. Such a curve is shown in figure 5 for the cell with 360° azimuth change. Similar effects in non-chiral nematics have been studied by Scheffer (1980) and Thurston (1982).

Plots of the configuration of the cell close to the operating voltage are useful in estimating the optical properties of the cell. The data used to plot the configuration may be used in an optics program to compute the optical properties in detail if an estimate is insufficient. Figure 6 shows the configurations of two stable states and the intermediate unstable state of the same bistable cell at 1.86 V. The outermost loop on the figure is the configuration of the cell with no applied field. The optical properties of the state with no field are very similar to those of the outermost of the bistable states because the director configuration is nearly the same. Figure 7 shows the three stable configurations of figure 6 pictorially, together with the alternative half-turn state.

LAMINAR FLUID DYNAMICS

In dynamic problems there is frequently a point where the director passes through the direction normal to the surfaces. The problem of evaluating β' in that case cannot be disposed of easily as for static solutions. In 1975 descriptions of two somewhat different ways to do computer simulation of twist cell dynamics, ignoring effects of mementum and compressibility, were published (van Doorn 1975; Berreman 1975 a). Van Doorn set up the Leslie–Ericksen torque equations (Ericksen 1961; Leslie 1968; Aslaksen 1971†) and used polar coordinates to describe director orientation when the director was not too nearly parallel to the polar axis. He used x- and y-components of the rectangular coordinates and a Lagrange multiplier when the director was close to the poles. The author used a non-orthogonal curved-net coordinate system that was limited in range mainly to the upper hemisphere. We have since taken van Doorn's idea of using rectangular coordinates and a Lagrange multiplier, but have done so in such a way that the multiplier is never explicitly evaluated, and the three coordinates appear on an equal footing in the equations. This method avoids changing coordinate systems and it also avoids limitations in range of rotation.

Shear flow complicates the problem but the method can be illustrated by considering only viscous resistance to local rotation. In that case the equations may be derived by using the same Helmholtz free-energy as in the statics problem and a Rayleigh dissipation function (see, for example, Goldstein 1950) defined by the expression

$$F_{\mathrm{d}} = \tfrac{1}{2}\gamma_1(\dot{n}_x^2 + \dot{n}_y^2 + \dot{n}_z^2).$$

A generalized Lagrangian function (neglecting kinetic energy) that allows the use of three Cartesian director components while restricting the director to fixed length is

$$L = -F_{\mathrm{H}} - \tfrac{1}{2}\lambda(n_x^2 + n_y^2 + n_z^2),$$

where λ is a Lagrange multiplier on the expression for the square of the constant director length. The equations of motion are

$$\frac{\partial L}{\partial n_i} - \frac{\mathrm{d}}{\mathrm{d}z}\left(\frac{\partial L}{\partial n_i'}\right) + \frac{\partial F_{\mathrm{d}}}{\partial \dot{n}_i} = 0.$$

† In Aslaksen (1971), sin should be replaced by cos in the middle of the matrix in equation (1.10), and r by $-\Gamma$ in equation (2.7).

When expanded, these three 'torque equations' have the form

$$Q_i + \lambda n_i = \gamma_1 \dot{n}_i.$$

The Lagrange multiplier may be eliminated by the following method. Multiply each of these three equations by the corresponding director component n_i and sum them to obtain

$$\sum Q_i n_i + \lambda \sum n_i^2 = \gamma_1 \sum n_i \dot{n}_i.$$

Then notice that $\sum n_i^2 = 1$ so that $\sum n_i \dot{n}_i = 0$. Hence

$$\lambda = -\sum Q_i n_i.$$

This expression for λ may be inserted in the three torque equations, which can finally be rewritten as

$$Q_i(n_j^2 + n_k^2) - n_i(n_j Q_j + n_k Q_k) = \gamma_1 \dot{n}_i.$$

The moving directors computed from these equations maintain unit total length to first order, but it is still necessary to avoid slow deviation by renormalizing them to unity at least occasionally. We do so after each step in time.

The equations just obtained will give the correct equilibrium configurations if allowed to run for a long time. Because shear flow and forces are ignored, the computed dynamic behaviour is somewhat slower than and different from the actual behaviour. However, the computations are faster than those that include flow. A version of this simplified relaxation method was used to find approximate dynamic and static equilibrium configurations (Baur *et al.* 1975) before we wrote the static configuration program.

We have not found a general Rayleigh dissipation function to describe shear flow. Instead we use two shear-force and three torque equations derived from the Leslie–Ericksen equations (Berreman 1975 *a*; Aslaksen 1971) when we include shear. We use the substitutions just described to subject the directors to the unit-length constraint without explicitly introducing the Lagrange multiplier into the three torque equations. The forms of the two shear equations are

$$\sigma_{zx} = R_{11} V'_x + R_{12} V'_y + \sum S_{1i} n_i$$

and

$$\sigma_{zy} = R_{21} V'_x + R_{22} V'_y + \sum S_{2i} n_i,$$

where V'_i is a component of shear velocity gradient. The spatially invariant shear force σ is adjusted to make the integral of V' equal zero in twist cells, or the relative velocity of the surfaces in shear experiments. The form of the three torque equations is

$$Q_i(n_j^2 + n_k^2) - n_i(n_j Q_j + n_k Q_k) = T_{i1} V'_x + T_{i2} V'_y + \gamma_1 \dot{n}_i.$$

The expansion of the terms Q, R, S and T in these five equations is lengthy but it may be inferred by referring to the detailed expansion into four equations in polar coordinates in Berreman (1975 *a*), where R, S and T were all included in T_{ij} and Q was represented by λ.

Ultimately the expressions for time derivatives of the directors are functions of the directors and their first and second derivatives with respect to z. To approximate the derivatives, some scheme that is equivalent to fitting a 'spline' curve through the level z in question and its near neighbours is used. The simplest spline is a parabola for each component of the director at one level, fitted at that level and its two nearest neighbours. Computations with this spline are fast and stable most of the time. However, when torques are dominated by shear flow the progress becomes weakly unstable, resulting in slow growth of a 'zig-zag' pattern in the director as a function of z. This instability is not changed by shortening the time steps. This effect can be avoided by smoothing the components once every 20 or 30 time steps with a spline consisting of a parabola fitted by least-squares among the five levels with two on each side of the adjusted level. When there is no such zig-zag instability, smoothing in this way proves to have almost no effect either on the equilibrium configurations or on the rate of flow or configuration change, even when it is done at every step. Once the zig-zag pattern is smoothed it seldom regains an appreciable amplitude in half a cycle of the operation of a twist cell. Even if the zig-zag pattern is uncontrolled it ultimately shrinks when the configuration nears the equilibrium curve, as shown in figure 8.

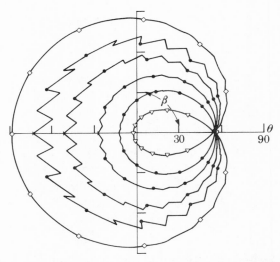

FIGURE 8. Second, field-free, half of a twist-cell cycle showing growing zig-zag pattern when smoothing is omitted. See figures 9 and 10 for other details, except that field remains off for 0.2 s for the outer loop in this figure.

All of the splines mentioned are simply weighted sums of each Cartesian component at the level in question and at one or two neighbours on either side. The weights needed to obtain the smoothed component and the first and second derivatives depend only on the distances to the neighbours. All these weights are computed at the start and stored in arrays for repeated use during the simulation of the dynamic process.

As with many relaxation programs this one breaks into violent temporal oscillation if time steps are too long. We do an approximate computation of the shortest period of oscillation or decay of any director if it is displaced while its two nearest neighbours are fixed. If we make the time steps shorter than this period divided by 2π we avoid the problem of oscillations.

RESULTS OF DYNAMIC SIMULATIONS

We have seen no measurements of the five viscosity parameters for E7. The viscosities of MBBA have been determined with fair accuracy. When we used those values in computations we obtained 'rise-times' from the DOWN state to the UP state in our bistable cells that were about half the observed time (Heffner & Berreman 1982). For lack of better information we shall assume that the viscosities of E7 are twice those for MBBA (see table 1). The numbers are not really in the right proportion because observed 'fall-times' from the UP state to the DOWN state are not quite as slow as these viscosities predict. Accurate values for five separate viscosities for the commonly used nematic mixtures would be very useful.

The possibility of simulating nematic fluid dynamics should allow for the design of simpler experiments to measure viscosities. The program we use allows one surface of the cell to move with respect to the other. With simulation to help in the interpretation, experimental cells with shear may be used to determine various combinations of viscosity parameters, depending on the orientation of the liquid crystal directors at the surfaces.

Even without accurate values for viscosity parameters it is instructive to follow a simulated cycle in a bistable twist cell. This is done in figures 9 and 10. The minimum time that the applied field must be held at a value other than the 'holding voltage' to cause a transition from one bistable state to another is just the time required for the configuration to pass beyond a configuration similar to the unstable equilibrium state between the two stable ones. From that time on, the elastic torques will carry the cell to equilibrium. However, by increasing the switching pulse times slightly, though not too much, the configuration will reach equilibrium in a minimum time because the state at the end of the pulse will closely resemble the intended equilibrium state. Such optimum times were used in figures 9 and 10.

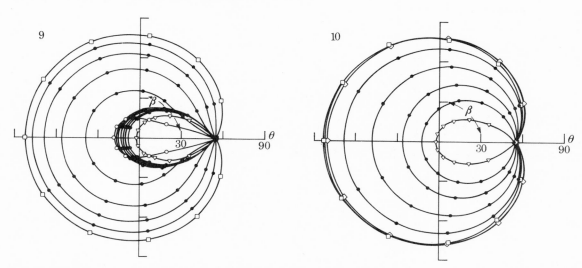

FIGURE 9. Dynamic progression from the equilibrium DOWN state of figure 2 (□), when thrice 1.86 V (5.62 V) is applied suddenly and continued for 0.01 s, after which the field reverts to 1.86 V. Successive configurations are 0.002 s apart with additional field on, 0.01 s apart with additional field off. Final configuration (▽) is near equilibrium UP state after 0.2 s. Viscosities are twice those for MBBA to give observed transition speed. The cell is 10 μm thick.

FIGURE 10. Progression from preceding UP configuration at 1.86 V (▽) after suddenly switching the field off for 0.1 s and then returning to 1.86 V. Successive configurations are 0.02 s apart. The measured 'down' transition is faster than computed with these viscosities.

Another use of the dynamics program has been to explain the movement of foreign particulate matter and disclinations to the edges of picture elements in ordinary twist cells (Berreman & Sussman 1979). Simulation of laminar flow shows a net movement of fluid across a picture element after repeated cycling. This will result in slower circulation around the element. Particles and disclinations that are weakly attached to the surface appear to be dislodged by the faster flow in the element and lodged at its downstream boundary.

An unexpected result of the simulation of flow in twist cells appears in the explanation of

[143]

214   D. W. BERREMAN

recent experiments of Hubbard & Bos (1981). They show that cells with a total twist of ¾ turn containing chiral nematics are fast and have reduced optical bounce. This result might seem to contradict previous indications that chirality slows the response of twist cells. However, Hubbard & Bos find that the same backflow effect that hinders return to the field-free configuration in ¼ turn cells, as indicated by van Doorn (1975) and Berreman (1975a), speeds that return in ¾ turn cells.

OPTICAL COMPUTATIONS

The 4 × 4 matrix method of optical computations has been adequately described by Smith (1965) and Berreman & Scheffer (1970). The method has been used in conjunction with a fluid dynamic program to show the complicated variations in twist-cell contrast with both time and viewing direction (Berreman 1975b). It has also been used to find optimal relations between cell thickness and optical anisotropy (Baur 1980, 1981; van Doorn et al. 1980; Birecki & Kahn 1980). Finally, we observe that in bistable twist cells the contrast is positive in some directions and negative in others. An example of this is shown in figure 11. As with ordinary twist cells, computer simulation has helped us to enhance the contrast and to understand how to improve the range of viewing directions.

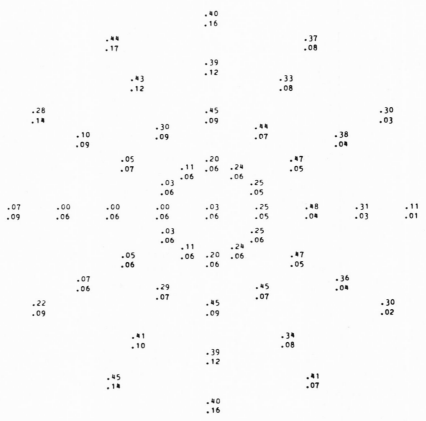

FIGURE 11. Transmission of plane-polarized light by cell similar to that of figure 6 but 16 μm thick, in DOWN (bottom numbers) and UP (top numbers) states, when viewed through a perfect crossed polarizer. Angles of view are normal and 10°, 20°, 30° and 40° from normal at 30° azimuth intervals. A region of negative contrast in one quadrant is characteristic of this cell, but it can be reduced in size and moved by altering parameters.

[144]

Modelling in more dimensions

The modelling alluded to so far is one-dimensional in the sense that director orientation is presumed to vary only along the z direction, normal to the cell surfaces. There are numerous problems that might be better understood through the simulation of configurations that vary in at least two dimensions. For example, lateral instabilities often occur in bistable or highly multiplexed twist cells in experimental stages of development. Greubel discovered this when he attempted to make a bistable, disclination-free chiral nematic twist cell with homeotropic boundary orientation (Greubel 1974). He then described a bistable cell with one state disordered by the instability.

Lateral variations of director configuration are crucial to 'dynamic-scattering' cells. Two- or three-dimensional simulation would also lead to better understanding of the fluid behaviour in such scattering cells.

In addition to lateral instabilities, there is the problem of lateral variation of director orientation and flow at the edges of picture elements in every twist cell. The effect of edges on static configurations is rather short-range; being on the order of the cell thickness. However, dynamic effects probably extend over larger distances.

Unfortunately even the analysis of two-dimensional variations is in a very rudimentary state of development. Analytic methods will probably always be restricted to small perturbations from laminar configurations unless inequalities in elastic constants are ignored. The complexity of boundary conditions necessitates the use of relaxation methods in computer simulation. So far, such computer simulations have only been used to find static configurations in the absence of fields (Berreman 1979). Electric field patterns would have to be adjusted after each change in director configuration unless the dielectric anisotropy is assumed to be too small to affect the pattern significantly.

Three-dimensional solutions have been studied even less. However, Sammon (1982) and Meiboom et al. (1983) have recently obtained static configurations that may simulate cholesterics in cubic 'blue phases' in the absence of fields. They used a fast relaxation method on a high-speed computer, but precision was low for reasonable running times.

The optics of liquid crystals with two- or three-dimensional variations in director is extremely complicated except in certain cases where ray optics or approximate scattering theories can be used. Fortunately, few cases have arisen where more detailed analysis seems necessary.

Conclusions

Despite present limitations to one dimension, numerical modelling has been very useful in understanding and optimizing the design of nematic and chiral nematic twist cells. If modelling were implemented in two or three dimensions with electric fields it would be possible to study many other important problems. Among these are stability against transverse distortions, movement of configuration walls and the effects of boundaries on flow.

I wish to acknowledge the hospitality and suggestions of the liquid crystal group at the Institut für Angewandte Festkörperphysik in Freiburg i.Br. during some of the development of the computer programs used in this work.

REFERENCES

Aslaksen, E. W. 1971 *Physik kondens. Materie* **14**, 80.

Baur, G. 1980 In *The physics and chemistry of liquid crystal devices* (ed. G. Sprokel), pp. 61–78. New York: Plenum Press.

Baur, G. 1981 *Molec. Cryst. liq. Cryst.* **63**, 45.

Baur, G., Windscheid, F. & Berreman, D. W. 1975 *Appl. Phys.* **8**, 101.

Berreman, D. W. 1975a *J. appl. Phys* **46**, 3746.

Berreman, D. W. 1975b In *Nonemissive electrooptic displays* (ed. A. R. Kmetz & F. K. Von Willisen), pp. 9–24. New York: Plenum Press.

Berreman, D. W. 1979 *J. Phys., Paris* **40** (suppl. 4), c3-58.

Berreman, D. W. 1980 In *The physics and chemistry of liquid crystal devices* (ed. G. Sprokel), pp. 1–13. New York: Plenum Press.

Berreman, D. W. & Heffner, W. R. 1982 In *S.I.D. Int. Symp. Digest Tech. Papers*, vol. 13, p. 242.

Berreman, D. W. & Scheffer, T. J. 1970 *Molec. Cryst. liq. Cryst.* **11**, 395.

Berreman, D. W. & Sussman, A. 1979 *J. appl. Phys.* **50**, 8016.

Birecki, H. & Kahn, F. J. 1980 In *The physics and chemistry of liquid crystal devices* (ed. G. Sprokel), pp 125–142. New York: Plenum Press.

de Gennes, P. G. 1975 *The physics of liquid crystals*, equation 6.43. Oxford University Press.

de Jeu, W. H. 1978 *Phys. Lett.* A **69**, 122.

Deuling, H. J. 1974 *Molec. Cryst. liq. Cryst.* **27**, 81.

Ericksen, J. L. 1961 *Trans. Soc. Rheol.* **5**, 23.

Frank, F. C. 1958 *Discuss. Faraday Soc.* **25**, 19.

Goldstein, H. 1950 *Classical mechanics*, chapter 1. Addison-Wesley Press.

Greubel, W. 1974 *Appl. Phys. Lett.* **25**, 5.

Heffner, W. R. & Berreman, D. W. 1982 *J. appl. Phys.* **53**, 8599.

Hubbard, R. L. & Bos, P. J. 1981 *IEEE Trans. Electron. Devices* **ED-28**, 723.

Leslie, F. M. 1968 *Arch. ration. Mech. Analysis* **28**, 265.

Leslie, F. M. 1970 *Molec. Cryst. liq. Cryst.* **12**, 57.

Meiboom, S., Sammon, M. & Brinkman, W. F. 1983 *Phys. Rev.* A **27**, 438.

Oseen, C. W. 1933 *Trans. Faraday Soc.* **29**, 883.

Raynes, E. P., Tough, R. J. A. & Davies, K. A. 1979 *Molec. Cryst. liq. Cryst. Lett.* **56**, 63.

Sammon, M. J. 1982 *Molec. Cryst. liq. Cryst.* **89**, 305.

Scheffer, T. J. 1980 In *Advances in liquid crystal research and applications* (ed. L. Bata), pp. 1145–1153. Oxford: Pergamon Press.

Smith, D. O. 1965 *Optica Acta* **12**, 13.

Thurston, R. N. 1982 *J. Phys., Paris* **43**, 117.

Thurston, R. N. & Berreman, D. W. 1981 *J. appl. Phys.* **52**, 508.

van Doorn, C. Z. 1975 *J. appl. Phys.* **46**, 3738.

van Doorn, C. Z., Gerritsma, C. J. & de Klerk, J. J. 1980 In *The physics and chemistry of liquid crystal devices* (ed. G. Sprokel), pp. 95–104. New York: Plenum Press.

Yang, K. H. 1982 *J. appl. Phys.* **53**, 6742.

Phil. Trans. R. Soc. Lond. A **309**, 217–229 (1983)
Printed in Great Britain

Liquid crystalline materials: physical properties and intermolecular interactions

By W. H. de Jeu

Solid State Physics Laboratory, Melkweg 1, 9718 EP Groningen, The Netherlands

The physical properties and the phase behaviour of some nematic liquid crystals are discussed, with emphasis on the influence of short-range antiparallel dipole correlation that occurs in mesogenic compounds with a strong terminal dipole moment. Evidence for this effect, which stems especially from dielectric studies, is summarized. Variations of the dipole correlation with molecular structure can explain the sometimes unexpected phase behaviour and physical properties of these substances. A qualitative model is given in terms of a monomer–dimer equilibrium.

1. Introduction

During the last decade a growing interest has developed in the 'molecular engineering' of liquid crystalline materials to obtain specific properties. Here three aspects can be distinguished: chemical properties (discussed by others at this meeting); phase behaviour, like mesomorphic temperature range and nematic–smectic behaviour; physical properties, like birefringence, dielectric anisotropy and visco-elastic properties. The demands with respect to the first two points being largely satisfied, attention has shifted to methods of influencing the physical properties. Here we can distinguish two points: the properties of the molecules themselves, and the interactions through which a molecular property influences the macroscopic behaviour.

So far attention has been directed mainly towards the first of these two. However, it seems that the interactions mentioned above are far more variable than was commonly thought some years ago. This insight is mainly due to new results from the synthetic chemists, who have presented us with new mesogenic compounds that often show a behaviour that is rather unexpected from a more traditional point of view. In this paper I wish to show that many of these unexpected results can be rationalized in terms of differences in short-range correlation between the molecules.

The plan of the paper is as follows. In § 2 some of the more classical relations between molecular properties and macroscopic physical properties will be summarized. Next, in § 3, the evidence for short-range correlations will be discussed, emphasizing results from relatively recently synthesized compounds with strongly polar end-groups like CN, saturated ring systems, or both. Finally these results are used to understand at least qualitatively some of the features of the observed phase behaviour (§ 4) and of the macroscopic physical properties (§ 5).

2. Mesogenic molecules and anisotropic physical properties

The orientational statistics of the molecules giving a nematic phase is in principle well known. Usually it is assumed that in spite of possible internal motions a rigid body can be used to represent a molecule in its 'average' conformation. The simplest choice is then a body with cylindrical symmetry, which implies that any possible bias in the rotation of the molecules around their long axis is disregarded. A molecular property can be described by a second-rank tensor $\boldsymbol{\lambda}$ which now has a particularly simple form: only the elements λ_1 (longitudinal) and λ_t (transverse) are different. The anisotropy observed macroscopically is related to $\lambda_1 - \lambda_t$ and to the degree of orientational order. The latter is given by $S \equiv \langle P_2 \rangle = \langle \frac{3}{2} \cos^2 \beta - \frac{1}{2} \rangle$, where the angular brackets indicate an average over all orientations, and β is the angle between the long molecular axis and the director n (preferred direction). I shall now discuss the type of relations given so far for various physical properties (de Jeu 1980, 1981).

(a) Birefringence

The birefringence of a nematic is given by the difference between the extraordinary and ordinary refractive index: $\Delta n = n_e - n_o = n_\parallel - n_\perp$, where the subscripts $\parallel$ and $\perp$ refer to the directions parallel and perpendicular to n, respectively. In practice Δn is positive and varies from values close to zero to about 0.4. The refractive index is related to the response of matter to an electric field. On application of a field E, an electric polarization P is induced, given by

$$P = \epsilon_0 (\epsilon - I) E, \tag{1}$$

where ϵ_0 is the permittivity of free space, ϵ the relative permittivity tensor, and I the unit tensor. Taking n parallel to the z axis of a laboratory frame, ϵ will be diagonal. In the optical frequency range

$$\epsilon_{ii} = n_i^2, \quad i = x, y, z, \tag{2}$$

with $n_\parallel = n_z$ and $n_\perp = [\frac{1}{2}(n_x^2 + n_y^2)]^{\frac{1}{2}}$. For a microscopic interpretation of the refractive index one considers the polarizability tensor α associated with a molecule. Then one arrives at a macroscopic polarization

$$P = N \langle \alpha \cdot E_i \rangle, \tag{3}$$

which can be equated to (1). Here N is the number of molecules and E_i the internal field, the sum of the macroscopic field and the average field of the dipoles of the surrounding molecules.

There are two reasons why attempts to predict the birefringence quantitatively are not very successful. In principle reasonable estimates of the electronic polarizabilities of a molecule can be obtained from the addition of tabulated bond polarizabilities. However, there is often an extra contribution, due to the conjugation of various groups, that is difficult to estimate. For mesogenic molecules this may be the most important contribution to the optical anisotropy. Secondly, there seems to be no satisfactory procedure to calculate the internal field and its possible anisotropy. In this situation the best one can do is inspect the molecule for the presence of unsaturated bonds and estimate where conjugation is possible. Comparison with results obtained from molecules with similar groups provides a fair idea of the possible birefringence.

(b) Dielectric permittivity

For non-polar compounds one finds for the static dielectric anisotropy similar values to those for $n_\parallel^2 - n_\perp^2$, as expected. By substitution of polar groups much larger anisotropies can be obtained. For p-cyano-substituted compounds anisotropies between 6 and 12 are found. When several polar groups are present, all having a dipole component in the same direction among the long molecular axis, $\Delta\epsilon$ values of more than 30 can be obtained. Substituting cyano groups at two *ortho* positions at a benzene ring, $\Delta\epsilon < -20$ has been observed.

In polar molecules the orientation polarization must be added to (3). Combination with (1) then gives

$$(\epsilon - I) E = (N/\epsilon_0) (\langle \alpha \cdot E_i \rangle + \langle \bar{\mu} \rangle), \tag{4}$$

where $\bar{\mu}$ is the average value of the permanent dipole in the presence of the electric field. For the problems related to the description of the internal field E_i and of the directing field involved in the calculation of $\langle\bar{\mu}\rangle$ we refer to the literature (Böttcher & Bordewijk 1978; de Jeu 1978). For strongly polar molecules the dipole contribution to ϵ dominates. Independently of the details of the theoretical treatment one finds

$$\bar{\epsilon} \approx \mu^2/T, \quad \Delta\epsilon \approx \mu^2 S/T. \tag{5}$$

As we shall see in the next section, in several cases deviations from these functional relations are observed that must be interpreted as being due to short-range correlation between the dipole moments. Consequently attempts to calculate $\Delta\epsilon$ from group moments (which are in principle well known) can easily be in error by a factor of two.

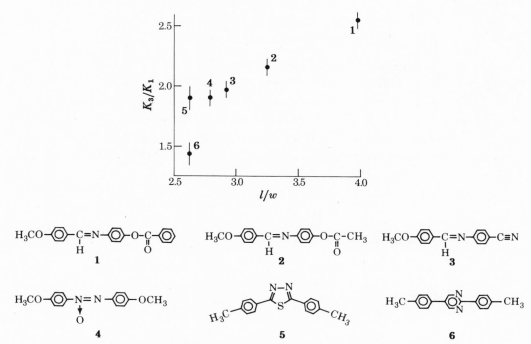

FIGURE 1. K_3/K_1 and the length/width ratio of some nematics without flexible alkyl chains (after Leenhouts & Dekker 1981).

(c) Elastic constants

The bulk elastic properties associated with the curvature of the director of a nematic liquid crystal are described by three elastic constants. These are associated with the restoring torques opposing splay (K_1), twist (K_2), or bend (K_3) of the director pattern. The ratio K_3/K_1 has been found to vary considerably, in contrast to K_2/K_1. For various compounds one finds approximately:

$$0.5 < K_3/K_1 < 4.0,$$
$$0.5 < K_2/K_1 < 1.2.$$

In fact in most cases $K_2/K_1 \lesssim 0.7$.

Leenhouts & Dekker (1981) established experimentally a relation between K_3/K_1 and the length/width ratio of the molecules. This is illustrated in figure 1, where also a new result for a banana-shaped molecule (5) has been included. In that case there is a possibility of adjusting to

[149]

the bend field, which could lead to a lowering of K_3. As we see, this effect, if any, must be small, as **5** fits rather well into the general trend for K_3/K_1. This relation with length/width ratio is roughly in agreement with theoretical calculations of various types. However, it holds experimentally only for relatively rigid molecules. Homologous series show a very different trend. Here, K_3/K_1 is found to *decrease* with increasing length of the alkyl chain. This is a quite dramatic effect. If for compound **1** the terminal benzene ring is replaced by a butyl chain (which does not differ much in length), K_3/K_1 decreases from 2.4 to 1.3. A model that accounts for these differences has recently been given by Van der Meer *et al.* (1982). It incorporates increasing smectic-like short-range order with increasing alkyl chain length. Within the framework of various approximations this can reproduce the trend observed experimentally. Because the elastic constants are related to the gradients of the molecular interactions they can indeed be expected to be rather sensitive to short-range correlations.

(d) Viscosity coefficients

There seems to be a renewed interest in the viscous behaviour of liquid crystals these days. From the point of view of applications one may wish to compare the viscosities of various classes of compound. Then a simple flow experiment in which the orientation of n is not controlled may suffice (Constant & Raynes 1980). As discussed by F. M. Leslie at this meeting, from a more fundamental point of view the situation is rather complicated. If in a simple shear experiment the orientation of n is fixed, three different viscosities are immediately evident (figure 2). In

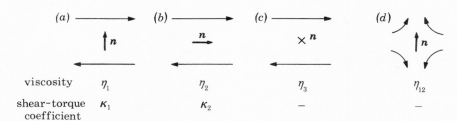

FIGURE 2. Definition of the various viscosity and shear-torque coefficients.

addition to these shears, which are antisymmetric in the coordinates, a symmetric contribution also exists (figure 2d). If the orientation of n is not fixed, the motion of the director comes also into play. In the cases of figure 2a, b the shear will exert a torque on n. The corresponding shear-torque coefficients (with the dimensions of a viscosity) are denoted by κ_1 and κ_2, respectively. Note that the associated rotation of n does not necessarily involve fluid motion. In the shear-plane in principle a situation of vanishing shear-torque can exist. Then n makes an angle θ_0 with the flow direction, given by

$$\tan^2 \theta_0 = -\kappa_2/\kappa_1. \tag{6}$$

Sorting out the precise temperature dependence of the nematic viscosities proves to be rather difficult, let alone the molecular influences. The viscosity of an isotropic liquid varies approximately as

$$\eta_{\text{is}} = \eta_0 \exp{(E/k_B T)}, \tag{7}$$

where $E > 0$ is an activation energy for diffusion. In the nematic phase the additional dependence on the order parameter will modify this activated process. As the temperature dependence of S is very weak compared with that in (7), this is difficult to determine. Kneppe *et al.* (1981) found that for several substances the reduced viscosities $\eta_i/\bar{\eta}$ $(i = 1, 2, 3)$ where $\bar{\eta} = \frac{1}{3}(\eta_1 + \eta_2 + \eta_3)$, behave in a very similar way (see figure 3).

As far as the shear-torque coefficients are concerned, many results are available for $\gamma_1 = \kappa_1 + \kappa_2$, but there is no agreement yet on the actual dependence on S (Prost *et al.* 1976; Diogo & Martins 1981). The coefficient κ_2 is often rather small, leading to values of θ_0 close to zero. Interestingly, in several cases κ_2 changes sign as a function of temperature, in which case there is no longer a real solution to (6).

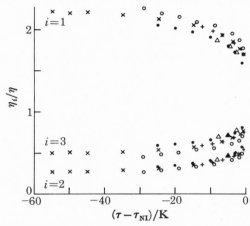

FIGURE 3. Reduced viscosities $\eta_i/\bar{\eta}$ for various compounds (after Kneppe *et al.* 1981).

3. Short-range correlations

During the last few years synthetic organic chemists have presented us with several series of new mesomorphic compounds. Two aspects are here of special interest:

extensive use of strongly polar end groups (especially CN) to obtain a large positive $\Delta\epsilon$ – the best known examples are the p,p'-alkylcyanobiphenyls (nCB series; see **7** in table 1);

replacement of the traditional aromatic rings (benzene) by saturated ring systems such as cyclohexane and bicyclo(2,2,2)octane.

Very often both aspects are found in combination, which is probably not accidental. Some examples of compounds with cyclohexane rings are given in table 1. We can compare with

$$C_7H_{15} - \bigcirc\!\!\!-\!\!\!\bigcirc - CN \qquad Cr \xrightarrow{61\,°C} N \xrightarrow{95\,°C} I$$

In the absence of a terminal CN-group, compounds with saturated ring systems show often pronounced smectic tendencies.

From table 1 we see that T_{NI} increases in the order **7**, **8**, **10**, and that T_{NI} of **9** is very low. *A priori* this order is somewhat unexpected: from the steric point of view these molecules are very similar, while one also does not expect the attractive dispersion forces to change in this order. Looking at some physical properties we note that n_{is} and also Δn decrease in the same order as T_{NI} increases. This decrease is evidently due to the replacement of the anisotropic and polarizable benzene rings by cyclohexane. More unexpectedly there are also important differences in ϵ_{is} at T_{NI}. This has often been too easily attributed to differences in conjugation. Conjugation of the CN dipole with an aromatic ring in fact can lead to two effects: an increase of the effective dipole moment, and delocalization of the dipole over the entire conjugated part of the molecule. The first effect is relatively small, as indicated in table 2, and is not sufficient to explain the large differences in ϵ_{is}. Furthermore ϵ_{is} is anyway smaller than expected for molecules with such a large dipole moment. This is all evidence for antiparallel dipole correlation. This idea is confirmed by the results given in figure 4, where we see that the temperature dependence of both ϵ_{is} and $\bar{\epsilon}$ in the nematic phase does not follow (5). In fact both are rather constant. This behaviour is typical for associated compounds, the association becoming stronger at lower temperatures. Then μ^2 must be replaced by $g\mu^2$, g being the dipole correlation factor. In our cases we find for **7**, **8** and **10** g of the order of 0.5.

TABLE 1. CLEARING POINTS AND SOME PHYSICAL PROPERTIES OF FOUR MESOGENS

	compound	m.p./°C	T_{NI}/°C	Δn†	$n_{is}(T_{NI})$	$\epsilon_{is}(T_{NI})$
7	C_7H_{15} ⟨◯⟩—⟨◯⟩— CN	29	42	0.16	1.56	9.7
8	C_7H_{15} ⟨⬡•⟩—⟨◯⟩— CN	30	57	0.09	1.50	8.5
9	C_7H_{15} ⟨◯⟩—⟨⬡•⟩— CN	18	−20†	—	—	—
10	C_7H_{15} ⟨⬡•⟩—⟨⬡•⟩— CN	71	83	0.05	1.47	5.6

† At $T_{NI} - T = 10$ °C and 589 nm.
‡ Approximate value extrapolated from binary mixtures.

TABLE 2. DIPOLE MOMENTS OF THE CYANO GROUP (MINKIN *et al.* 1970)

compound	$\mu_{gas}/10^{-30}$ C m	$\mu_{solution}/10^{-30}$ C m†
C_4H_9CN	13.6	11.9
⟨◯⟩— CN	14.6	13.1

† In benzene at 25 °C.

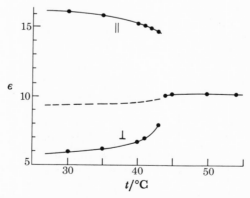

FIGURE 4. Static dielectric permittivities of 7CB (**7**).

Antiparallel dipole correlations in liquids can be described in a model as a monomer–dimer equilibrium:

$$D \underset{\rightarrow}{\overset{\leftarrow}{=}} 2M. \tag{8}$$

The associated pairs need not necessarily exist on a human time scale, and a dynamic equilibrium will be involved. From the permittivity data of the CB series one can calculate values of the order of 0.4 for the dimer concentration x_D^-. Higher order n-mers will occur only with a low probability. This is mainly due to the effect of frustration (Toulouse 1977): if three parallel molecules are put on, for example, a triangular lattice, two can orient their dipole moments antiparallel, but the direction of the third is undetermined (figure 5). Only very specific choices of the lattice can avoid this frustration effect, for example four dipoles on a square lattice. The important conclusion is that with spatial disorder, as in a liquid, frustration of n-mers cannot be avoided. Only the dimers are free of it, and the formation of these is the most efficient way to lower the electrostatic part of the free energy. Summarizing, we can say that the formation of dimers is a consequence of the antiferroelectric interactions between dipole moments and probably also induced dipoles, in combination with the liquid structure.

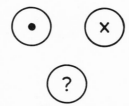

FIGURE 5. Frustration effect for three parallel dipoles on a triangular lattice.

The idea of molecular association of nitriles in both the nematic and the isotropic phase is supported by X-ray measurements (Leadbetter et al. 1975) and nuclear magnetic resonance studies (Saito et al. 1973). Additional information can sometimes be obtained from the layer spacing of a smectic A phase, when this phase occurs at lower temperatures. The spacing is usually not commensurate with the molecular length l, but varies for different compounds from $1.1 l$ up to $1.6 l$. This has been interpreted as variations in the degree of overlap of the molecules forming the dimer.

4. APPLICATIONS TO PHASE BEHAVIOUR

(a) Nematic–isotropic phase transition

According to the van der Waals theory of nematics, discussed by M.A. Cotter at this meeting, the isotropic–nematic phase transition is due to the combination of attractive dispersion forces and an anisotropic excluded volume. The new insight from it is that the dispersion forces themselves need not be anisotropic. An orienting effect is obtained anyway when combined with an anisotropic excluded volume. Applied to mesogenic molecules in a very qualitative way, one could say that T_{NI} increases with: (i) increasing molecular polarizability, and (ii) an increasingly anisotropic excluded volume. This seems to fit in with observations on many molecules, and the better that one condition is fulfilled, the less the other is needed.

Combination of these ideas with the monomer–dimer equilibrium could provide the explanation for some of the unexpectedly high T_{NI} values, as already noted by Gray (1981). The degree of overlap of two monomers when forming a dimer depends on the extension of the permanent dipole moment (related to the amount of conjugation), and on the possibilities of creating

[153]

induced dipole moments in the most polarizable parts of the molecule. Often, but not necessarily, these two effects combine. In table 3 I have drawn the structures of the dimers for the compounds of table 1, as they follow from this point of view. In the order 7, 8, 10 there is a decrease in polarizability (both for monomer and dimer), very little change in the excluded volume of the monomers, and a strong increase of the anisotropy of the excluded volume of the dimers. Provided that the monomer–dimer equilibrium does not shift much, the latter effect might dominate, which explains the order of T_{NI}. Compound 9 is very different from the others in that it has a strongly polarizable benzene ring separated from the strong CN dipole by a saturated ring. This allows for two types of dimers. Of these 9b can be expected to be more stable because of the added interactions at two 'contact points' (Vorländer). But with respect to the anisotropy of the excluded volume 9b differs little from 7. Then the decrease in polarizability compared with 7 will dominate, leading to a decrease of T_{NI}.

TABLE 3. POSSIBLE TYPES OF DIMER WITH THE COMPOUNDS OF TABLE 1

The explanation given for the order of T_{NI} of compounds 7–10 is rather qualitative, and several points should be made more precise. In particular the equilibrium constants of the monomer–dimer equilibrium should be determined from careful measurements of the permittivities over a wide temperature range. Such measurements are in progress in Leuven (Belgium). However, to the best of my knowledge no other explanation is yet available for the rather unexpected order of the clearing points of these four compounds. Quite generally it has been found that separation of polar or polarizable groups in a molecule (benzene, cyano, oxygen with its lone-pair electrons) by a saturated ring or group, tends to reduce T_{NI}. Perhaps this can again be attributed to an unfavourable short-range structure, due to the possibility of interactions at two 'contact points'.

(b) Re-entrant nematic behaviour

Another aspect of the association between strongly polar molecules is the possible occurrence with decreasing temperature of the phase sequence nematic–smectic A–re-entrant nematic. This was first observed by Cladis (1975) in a mixture of two *p*-cyano-substituted compounds. Later it was also observed in a pure compound at high pressure and finally even in pure compounds at atmospheric pressure. Recently a microscopical model has been proposed for these two subsequent phase transitions, one that makes explicit use of the monomer–dimer equilibrium (Longa & de Jeu 1982).

The first phase transition nematic–smectic A can be understood from a simple extension of McMillan's theory of the smectic A phase. In his treatment the smectic layering is stabilized if a quantity r_0/l is small enough, where r_0 is the length of the central aromatic core (which is assumed to be strongly polarizable) and l the total length of the molecule. In its original form this applies only to approximately symmetrical molecules. With increasing molecular asymmetry the smectic phase can be expected to be destabilized. As a consequence the monomers of, for example, the cyanobiphenyls will have only weak smectic tendencies. However, if pairing occurs, the dimers are again symmetric and can fulfil the condition of small r_0/l. Hence a smectic phase can be expected if there is (1) sufficient pairing (low temperatures or high pressures), together with (2) long alkyl chains (small r_0/l of the dimers). The phase transition is percolation-like: above a certain threshold concentration of dimers the smectic order becomes long-range.

Interestingly the model calculations show that from the point of view of packing, the monomers help to stabilize the smectic A phase with a layer spacing equal to the length of the dimer. In the presence of monomers, space is better filled than when a dimer is surrounded by only dimers. The smectic order parameter increases with decreasing temperature or increasing pressure. At the same time the equilibrium shifts to the dimer side. At a certain stage this will cause an increasingly difficult packing of the bulky dimers in the smectic planes, while less and less monomers are available to fill the outer regions of the smectic layers. Because of the unfavourable packing entropy, dimers then have to move out of the smectic planes. The smectic order parameter thus decreases and finally the nematic phase re-enters.

Independently of model parameters re-entrant behaviour is predicted to be strongly favoured for length ratios of dimer and monomer of the order of 1.3–1.5. This is in agreement with experiments, where layer spacings of this magnitude are often found. For dimers of type **10** (table 3) this ratio is larger, and indeed for these types of compound no re-entrant behaviour has been observed so far.

5. APPLICATIONS TO PHYSICAL PROPERTIES

Dielectric studies are an important means of obtaining information about molecular association. Well studied are the CB series (Dumnur & Miller 1980; Parneix 1982). For a more quantitative approach results for the permittivity over a wide temperature range are required. This means necessarily measurements high up into the isotropic phase until the temperature dependence is close to (5). Though the monomer–dimer equilibrium will be determined mainly by the short-range structure, a small discontinuity in $\bar{\epsilon}$ at T_{NI} is well known. Moreover some dependence of $\bar{\epsilon}$ in the nematic phase on the order parameter can be observed close to T_{NI} (Schad & Osman 1981). The dynamics of the permittivity might provide additional information (Wacrenier *et al.* 1981).

The importance of differences in short-range correlations can be demonstrated once more with the example of compound **11**.

11

when X is benzene, $\epsilon_{\parallel} \approx 35$, $\epsilon_{\perp} = 7$; when X is cyclohexane, $\epsilon_{\parallel} \approx 20$, $\epsilon_{\perp} = 7$. In spite of the similar group moments there is a large difference in $\epsilon_{\parallel}$ of the two molecules. This example also illustrates the problems if we want to interpret the permittivity results quantitatively. In the first place there is a difference in conjugation between the two molecules, leading to somewhat different total dipole moments. Secondly one can expect a difference in short-range antiparallel correlation. The first effect should be exactly known to make the second one more quantitative.

In connection with applications, much work has been done lately on the elastic properties of various classes of compounds. There is interest especially in K_3/K_1, which should be low for a steep threshold of twisted nematic displays. Because of the required positive $\Delta\epsilon$, compounds with a terminal CN group are again often considered. Comparison of the results for various compounds is sometimes hindered by the fact that the same alkyl chain length cannot always be used. This mixes up the two trends described in § 2 c. Nevertheless, even if this is taken into account it is clear that the simple relation between K_3/K_1 and length/width of the molecules does not hold generally. In particular two effects have been observed.

(i) Replacements of a benzene ring by a saturated ring tends to increase K_3/K_1 (Bradshaw & Raynes 1981; Schad & Osman 1981; Schadt & Gerber 1982). A bicyclo(2,2,2)octane ring has a more pronounced effect than cyclohexane (Bradshaw *et al.* 1981).

(ii) Introduction of heterocyclic rings (whether saturated or not) tends to decrease K_3/K_1 (Scheuble *et al.* 1981; Schadt & Gerber 1981; see also **6** in figure 1).

Schad & Osman (1981) have tried to explain the first point by incorporating the influence of the dimers on the effective length/width ratio. For compounds **7, 8** and **10** the length/width ratio of the monomer is about the same, while that of the dimer increases in the order given. Hence, provided that the number of dimers does not decrease strongly in this order, the results for K_3/K_1 then fit in with Leenhouts's general trend. The second effect could in principle be rationalized along similar lines. For the systems investigated the hetero atoms give an extra dipole moment that adds up with the cyano dipole. This could increase the degree of dimerization. Taking the relatively unfavourable leng/thwidth ratio of the dimers (compared with the monomers) into account, this lowers K_3/K_1. However, at this stage I feel that explanation and speculation are no longer very well distinguished.

In conclusion it still remains to be seen whether the relation between K_3/K_1 and length/width of the molecule can be saved by incorporating the short-range correlations described above. More importantly the other trend of K_3/K_1 decreasing with increasing alkyl chain length is still not fully understood.

Coming finally to the viscosities one can point out two effects that are probably related to short-range correlations and that would be worth further investigation. In the first place for various compounds some important differences have been observed in the activation energy, which governs the temperature dependence (equation (7)). This is again somewhat unexpected for molecules of similar size and with similar clearing points. However, if a monomer–dimer equilibrium is involved it would be very natural that the viscosity is also influenced by this equilibrium. Secondly, the flow alignment angle θ_0 (see equation (6)) is usually small. The few substances for which κ_2 is known to change sign (see figure 6) are again compounds with a terminal CN group. In fact Gähwiller (1972) speculated already that the special shape of the dimer is responsible for this effect. At present we are comparing in Groningen the flow alignment of compounds with and without a CN group, that are otherwise as similar as possible, to verify whether this is indeed so.

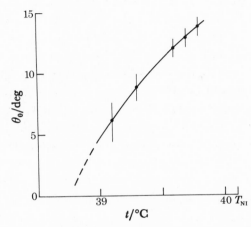

FIGURE 6. Region of existence of the flow alignment angle for 8CB
(after Beens & de Jeu 1983).

In this paper I have tried to show that for mesogenic compounds with a strong terminal dipole moment, short-range antiparallel dipole correlation is of decisive influence both for the phase behaviour and for the various physical properties. It is somewhat disappointing that this cannot be made more quantitative yet. A key question is whether the monomer–dimer description is valid. However, if one chooses not to believe in dimers, it is possible to reformulate the various arguments in terms of direct correlation functions. To obtain quantitative information on these will then be as difficult as to describe the monomer–dimer equilibrium more precisely.

REFERENCES

Beens, W. W. & de Jeu, W. H. 1983 *J. Phys.* (In the press.)
Böttcher, C. J. F. & Bordewijk, P. 1978 *Theory of electric polarization*, 2nd edn, vol. 2. Amsterdam: Elsevier.
Bradshaw, M. J., McDonnell, D. G. & Raynes, E. P. 1981 *Molec. Cryst. liq. Cryst.* **70**, 289–300.
Bradshaw, M. J. & Raynes, E. P. 1981 *Molec. Cryst. liq. Cryst. Lett.* **72**, 35–42.
Cladis, P. E. 1975 *Phys. Rev. Lett.* **35**, 48–51.
Constant, J. & Raynes, E. P. 1980 *Molec. Cryst. liq. Cryst.* **62**, 115–124.
Diogo, A. C. & Martins, A. F. 1981 *Molec. Cryst. liq. Cryst.* **66**, 133–146.
Dunmur, D. A. & Miller, W. H. 1980 *Molec. Cryst. liq. Cryst.* **60**, 281–292.
Gähwiller, C. 1972 *Phys. Rev. Lett.* **28**, 1554–1556.
Gray, G. W. 1981 *Molec. Cryst. liq. Cryst.* **63**, 3–18.
de Jeu, W. H. 1978 In *Liquid crystals* (ed. L. Liebert) (*Solid State Suppl.* no. 14), pp. 109–145. New York: Academic Press.
de Jeu, W. H. 1980 *Physical properties of liquid crystalline materials.* New York: Gordon & Breach.
de Jeu, W. H. 1981 *Molec. Cryst. liq. Cryst.* **63**, 83–110.
Kneppe, H., Schneider, F. & Sharma, N. K. 1981 *Ber. BunsenGes. phys. Chem.* **85**, 784–789.
Leadbetter, A. J., Richardson, R. M. & Colling, C. N. 1975 *J. Phys., Paris (Colloq.)* **36**, C1-37–C1-43.
Leenhouts, F. & Dekker, A. J. 1981 *J. chem. Phys.* **74**, 1956–1965.

228 W. H. DE JEU

Longa, L. & de Jeu, W. H. 1981 *Phys. Rev.* A **26**, 1632–1647.
Van der Meer, B. W., Postma, F., Dekker, A. J. & de Jeu, W. H. 1982 *Molec. Phys.* **45**, 1227–1243.
Minkin, V. I., Osipov, O. A. & Zhdanov, Yu. 1970 *Dipole moments in organic chemistry*, p. 67. New York: Plenum.
Parneix, J. P. 1982 Ph.D. thesis, University of Lille.
Prost, J., Sigaud, G. & Regaya, B. 1976 *J. Phys. Lett.* **37**, L341–L343.
Saito, H., Tanaka, Y., Nagata, S. & Nukada, K. 1973 *Can. J. Chem.* **51**, 2118–2133.
Schad, H. & Osman, M. A. 1981 *J. chem. Phys.* **75**, 880–885.
Schadt, M. & Gerber, P. R. 1982 *Z. Naturf.* **37a**, 165–178.
Scheuble, B. S., Baur, G. & Meier, G. 1981 *Molec. Cryst. liq. Cryst.* **68**, 57–67.
Toulouse, G. 1977 *Communs Phys.* **2**, 115–119.
Wacrenier, J. M., Druon, C. & Lippens, D. 1981 *Molec. Phys.* **43**, 97–112.

Discussion

E. P. RAYNES (*R.S.R.E., Malvern, U.K.*). I have two comments on Dr de Jeu's attempt to relate clearing point to the extent of dimerization.

(1) His discussion considered only the geometry of the dimer; I would have thought that the strength of the dimer (in other words the value of g) would be at least equally important.

(2) Evidence from X-rays and electric permittivities indicates that the antiparallel ordering (dimerization) in cyano nematics is reduced to negligible proportions by adding non-cyano material (for example dialkyl esters). However, contrary to your model there is no corresponding decrease observed in the clearing point of the mixtures.

W. H. DE JEU. A full quantitative model, if one were possible, would require a description of the equilibrium process monomer–dimer, in which the strength of the dimer is an important factor. I agree that in principle dilution studies could provide valuable tests of the ideas presented, but the interpretation of the results might be not as simple as suggested. Adding a non-cyano material to cyano nematics can do more than just breaking up the dimerization of the cyano compound. Often induced smectic phases are observed, probably indicating specific interactions between the cyano compound and the other one.

A. H. PRICE (*Edward Davies Chemical Laboratories, University College of Wales, Aberystwyth, U.K.*). Two major points of interest arise from this paper. The first concerns the assumption regarding the predominance of dimers in the nematic phase. The dielectric evidence for molecular association arises from the Kirkwood g factor being less than unity (as is found in some 'normal' polar liquids such as nitrobenzene) but this only tells us that antiparallel dipolar association exists and very little about the number of dipoles involved. From the results of Kirkwood g factor calculations one would not conclude that nitrobenzene associates into 'dimers'. Why do so for the nematogens mentioned?

The second point concerns the relative magnitude of dipole–dipole and dipole–induced-dipole interactions. The forces responsible for molecular association are not easily evaluated and the resulting molecular configuration of minimum potential energy is a balance between all the forces involved. Calculations of the magnitude of dipole–dipole and dipole–induced-dipole interactions invariably show that the former dominate. What evidence is available that this does not occur in the nematogens discussed in this presentation?

W. H. DE JEU. The g factor as obtained from permittivity measurements can be interpreted either in terms of direct correlation functions or as a monomer–dimer equilibrium. In strongly

[158]

terminally polar liquid crystals the occurrence of a smectic A phase with $d \approx 1.4\,l$ leads to some preference for considering an associated pair of molecules as a separate entity. This concept has been very useful in a molecular model of re-entrant nematic behaviour.

I quite agree that the various contributions to the intermolecular forces are not easily evaluated. The main point I wanted to make is that for the systems under consideration here, often emphasize has been on the interactions between permanent dipoles, overlooking possible interactions between dipole and induced dipole.

M. G. CLARK (*R.S.R.E., Malvern, Worcs., U.K.*). It is well known that the coefficient κ_2, which is more commonly denoted α_3, changes from a small negative value through zero to a diverging positive value as one approaches a smectic A phase through a nematic phase. Furthermore, studies of flow alignment instability during oscillatory shear (M. G. Clark, F. C. Saunders, I. A. Shanks & F. M. Leslie, *Molec. Cryst. liq. Cryst.* **70**, 195–222 (1981)) have shown that κ_2 may also become positive in materials without smectic phases. Indeed, to our knowledge all 'two-frequency' nematics (having the lowest relaxation of $\epsilon_\parallel$ at audio frequencies) have κ_2 positive over part or all of their nematic range. Thus the phenomenon of κ_2 changing sign is actually quite common and occurs in compounds without terminal CN groups as well as those possessing them.

Phil. Trans. R. Soc. Lond. A **309**, 231–238 (1983)
Printed in Great Britain

Materials for applications: the industrial chemists' viewpoint

By B. Sturgeon

BDH Chemicals Limited, Broom Road, Poole, Dorset, BH 12 4NN, U.K.

Many of the advances in displays during the past decade have been facilitated by the development of new mixtures, tailored to the application, in chemical companies that specialize in the production of liquid crystals. Problems that occur in this type of work are illustrated by the example of the development of mixtures for multiplexing 16–32 ways. The production of liquid crystals is discussed briefly.

Introduction

Companies that manufacture liquid crystals have made considerable contributions to the science of the subject and to the progress of displays. Merck began their work in 1968 at a time when few practical applications were known, and the materials they provided for experimental purposes helped to sustain the interest of research workers. The discovery that founded the display industry, the principle of the twisted nematic cell, was made in 1971 at Hoffmann La Roche and the cyanobiphenyls produced in 1973 were the first materials that made the production of low-voltage fast-switching displays possible. A decade of rapid progress has followed in which the succession of demands from the display industry for materials suitable for more sophisticated applications has been met by research into the chemistry and physics of liquid crystals. From the simplest static-driven watch display, the technology has advanced through three-way multiplexed displays for calculators, dot matrix displays, displays for operating over wide temperature ranges, coloured displays to the verge of high-information v.d.u.-type displays. Each application requires a liquid with different properties and each has been met from the products produced by the materials companies.

These developments occurred outside the display industry because companies with the capability to manufacture displays had little reason, before the l.c.d., to employ the physical and organic chemists necessary for research into liquid crystals, and even less incentive to invest in the chemical plant required for their manufacture. This situation remains largely unchanged today. Although there are some exceptions, most of the display industry continues to purchase its liquid crystal requirements. There is a natural industrial and technological boundary between the chemical companies that manufacture liquid crystals and the electronic companies that manufacture displays, and this gives the materials companies the opportunity to develop and sell their products.

Liquid crystals are examples of chemicals that the chemical industry describes as 'speciality chemicals', i.e. chemicals produced to be sold to a different industry for the sake of their properties rather than their structure. Speciality chemicals are attractive to the chemical industry because the prices they command are higher than those of commodity chemicals. On the other hand, their development is much more expensive because the chemical company must employ a group of experts in the speciality field that their business might not otherwise need. The expertise that these groups build up by their research, protected by patents, deters other chemical companies from competing in the business, and helps to preserve the position of those first in the field.

The sequence of events leading to the development and marketing of a new liquid crystal mixture forms a cycle of interactions illustrated in figure 1. The process may begin with a request from a customer for an improved material. More often it starts with the synthesis of a new chemical structure, and a hint, obtained from physical measurements, of properties superior to existing materials. In either case, the group concerned with making liquid crystal mixtures becomes involved, and if the efforts are successful the product is produced and marketed after suitable processes for manufacturing the chemicals have been devised. The research work of the

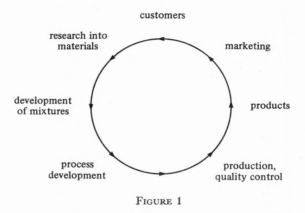

FIGURE 1

materials companies on one side of the circle is well known from their papers and patents (and here we at BDH are very fortunate to be associated with work at Hull University and the Royal Signals and Radar Establishment) and their products are well known on the other. The segment in between is the subject of this paper.

THE PREPARATION OF LIQUID CRYSTAL MIXTURES

The properties that the device engineer seeks are shown in table 1. One property may be of dominating importance for a particular application – such as fast response times at low temperatures for displays to be used for automobile instrumentation – but other criteria always have to be fulfilled as well. The scientist in the materials company must build the appropriate combination of properties into the mixture.

TABLE 1. PROPERTIES REQUIRED IN MIXTURES FOR DISPLAY APPLICATIONS

temperature range	threshold voltage
birefringence	sharpness of threshold
response times	temperature dependence of threshold

The process of making mixtures is guided by theoretical considerations, by calculation and by prediction, but much remains to be determined experimentally. The measurement of dielectric anisotropy and elastic constants on individual components is a valuable guide to which to use for a particular application or how best to exploit the properties of a new material. However, such measurements never provide the complete answer to making a mixture because components will have to be used that have transition temperatures inconveniently high for the measurements to be made, and the properties of individual components do not add together in a predictable manner in the mixture. Moreover, non-ideal behaviour may occur in the mixture, such as the formation of an injected smectic phase when nematics of high positive and low positive or negative dielectric anisotropy are mixed.

In the example described in detail below, the development of mixtures for multiplexing 16–32 ways, the voltages provided by the electronic circuit that will drive the display determine the performance required of the liquid crystal. This subject was discussed by Needham (this symposium), who showed that the ratio V_{ON}/V_{OFF} varies from about 1.3 for 16-way to 1.2 for 32-way multiplexing. She also showed that these voltages can be related to the performance of the liquid crystal via its electro-optic transmission curves and that $V_{50,10,20}/V_{90,45,20} = M'_{20} < V_{ON}/V_{OFF} < 1.3$ for 16-way or 1.2 for 32-way multiplexing when the voltages are measured in transmission. The voltage margin within which the liquid crystal must switch ($V_{ON} - V_{OFF}$) decreases with increasing level of multiplexing but increases with increasing supply voltage. Calculation of the voltage margin for 16–32-way multiplexing shows that the desired performance will be obtained only in mixtures with threshold voltage in the range 2–4 V, that is mixtures with a relatively low dielectric anisotropy. Clearly the threshold must be as sharp as possible.

Raynes *et al.* (1980) have shown that the sharpness of the capacitance threshold is inversely proportional to

$$\Delta\epsilon/\epsilon_\perp + \tfrac{5}{8}k_{33}/k_{11},$$

and a similar relation almost certainly applies to the optical threshold. The fraction $\Delta\epsilon/\epsilon_\perp$ is not a fruitful source for manipulation by varying the composition because $\Delta\epsilon$ is also a major factor in determining the threshold voltage and will be determined conversely by the necessity of keeping this in the range 2–4 V, whereas $\epsilon_\perp$ is decided mainly by the structure of the core of the molecule and is almost constant at 4–5 in common liquid crystals unless a lateral substituent is present. On the other hand, the ratio of elastic constants, k_{33}/k_{11} is known to decrease with increasing chain length in several homologous series of liquid crystals.

Thus analysis and calculation tell us that the materials most likely to be suitable for high-level multiplexing will have low dielectric anisotropy with a threshold in the range 2–4 V, with long chains at the end of the molecules to produce a favourable ratio of elastic constants and the sharpest possible optical threshold. Practical considerations add another constraint, in that it is advantageous to offer the device engineer two mixtures miscible in all proportions but with different threshold voltages so that he can tune the threshold exactly to match his cell and electronics.

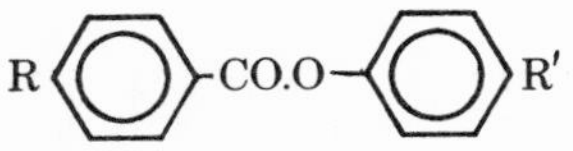

FIGURE 2. Structure of 4-alkylphenyl-4'-alkylbenzoic acid esters.

The class of liquid crystals chosen to fulfil these requirements was the 4-alkylphenyl-4'-alkylbenzoates (figure 2) because they are stable, have the lowest viscosity and highest birefringence among similar materials, are cheap to produce and have a proven record of performance in commercial liquid crystal mixtures. Their names are abbreviated to MEXY where X is the number of carbon atoms in the alkyl side chain attached to the acid part of the molecule and Y represents similarly the chain attached to the phenol.

The observation that started the development of these mixtures was made by D. G. McDonnell at R.S.R.E., who found that ME77 mixed with 5 % 2CB had an extremely sharp threshold (see table 2). This mixture froze at 22.5 °C, was nematic to 32 °C and had a monotropic transition to a smectic phase at 17 °C. The problem was to expand the nematic range to − 10 to 60 °C while retaining the key property: the sharpness of the threshold.

TABLE 2. ELECTROOPTIC DATA FOR MIXTURES OF HIGH POSITIVE $\Delta\epsilon$ COMPONENTS WITH MIXTURES OF ESTERS

	positive $\Delta\epsilon$ component	ester mixture	$\dfrac{V_{90,0,20}}{V\text{ r.m.s.}}$	M_{20}	M'_{20}
1	5 % 2CB	ME77	3.73	1.49	1.19
2	5 % 2CB	TE2	2.83	1.57	1.22
3	5 % 2CB	80 % TE2, 20 % MB75F	3.85	1.53	1.20
4	5 % 2CB	72 % TE2, 18 % MB75F, 10 % 3R33	3.67	1.56	1.22
5	12 % 2CB 3 % BB21	72 % TE2, 18 % MB75F, 10 % 3R33	2.48	1.61	1.27
6	2 % 2CB 3 % BB21	72 % TE2, 18 % MB75F, 10 % 3R33	3.89	1.55	1.22

The first property to receive attention was the freezing point, which is influenced strongly by the freezing point of the ester as the major constituent of the mixture. Other esters of a similar chain length had equally unfavourable melting points, and the only remaining possibility to depress the freezing point was to use a eutectic mixture of esters. At once the kind of compromise with which this type of work is fraught was encountered. Depression of melting point between homologues of similar chain length is generally small. Maximum depressions are obtained by maximum differences in chain length, and, because the chain cannot be lengthened beyond 7–8 carbon atoms owing to the occurrence of smectic phases, homologues of shorter chain length have to be used, which probably have less favourable elastic constants. The eutectic (TE2) of ME73, ME75 and ME77 was determined and found to freeze at − 3 °C and be nematic to 28 °C. Electrooptic measurements on a mixture of this with 5 % 2CB (see table 1) show there has been some loss of sharpness, but shortening the average chain length has also reduced the threshold voltage.

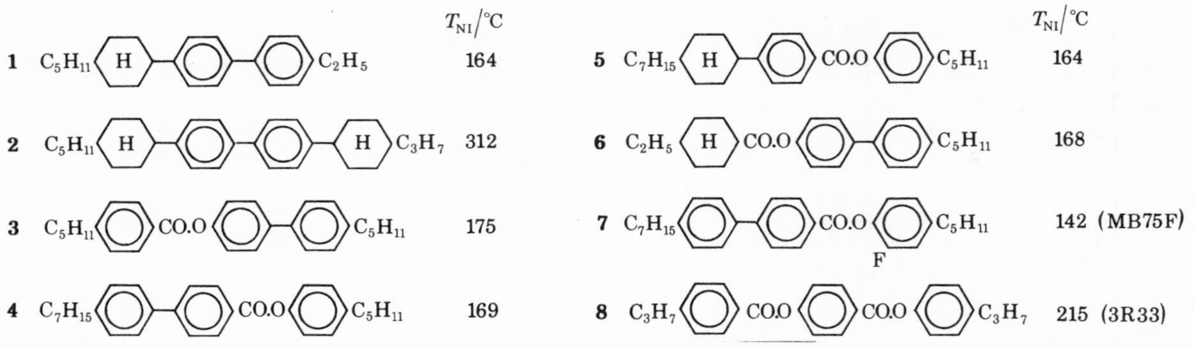

FIGURE 3. Low $\Delta\epsilon$, high T_{NI} components.

The next problem was to raise the N–I temperature (T_{NI}) to about 60 °C, for which purpose a component with low dielectric anisotropy but high T_{NI} was required. Several were studied (figure 3). The hydrocarbon molecules 1 and 2 were insufficiently soluble. The esters 3–6 either increased the viscosity too much or spoiled the threshold sharpness. Only MB75F and 3R33 seemed promising. The solubility of MB75F in TE2 was now determined and found to be about 20 % (figure 4), depressing the melting point to − 8 °C and raising T_{NI} to 47 °C. 3R33 was found similarly to be about 10 % soluble in 80 % TE2 – 20 % MB75F, depressing the freezing point further to − 10 °C, and raising T_{NI} to 58 °C. Measurement of the threshold on these mixtures after adding 5 % 2CB showed that little further deterioration in sharpness had occurred (see table 2).

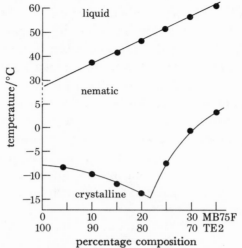

FIGURE 4. Part of the phase diagram of TE2 and MB75F.

TABLE 3. S–N TRANSITION TEMPERATURES OF 5 % OF HIGH POSITIVE Δε COMPONENTS
WITH 95 % OF A MIXTURE OF 72 % TE2, 18 % MB75F AND 10 % 3R33

positive Δε component	abbreviation	S–N temperature °C
C_2H_5 ⬡—⬡ CN	2CB	−7.4
C_5H_{11} ⬡—⬡ CN	5CB	+11.1
C_7H_{15} ⬡—⬡ CN	7CB	+16.7
C_3H_7 ⬡ H —⬡ CN	PCH3	+5.7
C_5H_{11} ⬡ H —⬡—⬡ CN	BICH 5	+25.2
C_2H_5 ⬡ H CO.O —⬡—⬡ CN	CHE	+15.2
C_7H_{15} ⬡—⬡ CO.O —⬡—⬡ CN	BB21	−13.6

[165]

Attention was next directed towards conferring on the mixture the positive dielectric anisotropy necessary to produce threshold voltages in the required range. Here positive materials with long chains are desirable (such as 7CB) for maximum benefit to the elastic constant ratio, and several were tested at 5 % by mass for their effect in generating injected smectic phases (see table 3). 2CB and BB21 stand out as best, perhaps because of their disparate lengths in comparison with the smectic layers. By adding the appropriate combination of these, two mixtures with all the required properties were obtained (see examples 5 and 6 in table 2). Unfortunately when these were then mixed together as intended by the user, an injected smectic phase was found to occur at intermediate compositions. Therefore the last resource to suppress such phases had to be deployed, which is to add a small quantity of a material with short chains at each end: 5 % ME15 in this case. The resulting mixtures, E130 and E140, could now be mixed together without producing a smectic phase stable above $-15\,^{\circ}\mathrm{C}$, and their properties, some of which were measured by E. P. Raynes and M. J. Bradshaw at R.S.R.E. are shown in table 4. E130 is suitable for 16-way multiplexing and E140 for 32-way, allowing for the increase in sharpness observed in reflexion in cells with less than 90° twist (Needham, this symposium).

TABLE 4. PROPERTIES OF E130 AND E140

	E130	E140
$T_{\mathrm{NI}}/^{\circ}\mathrm{C}$	64.3	63.0
$T_{\mathrm{SN}}/^{\circ}\mathrm{C}$	< -20	< -20
birefringence (at 589.6 nm, 20 °C)	0.154	0.147
$V_{90,0,20}/\mathrm{V}$	2.38	3.80
M_{20}	1.63	1.58
M'_{20}	1.27	1.22
$\epsilon_{\perp}$ at 20 °C	4.57	4.19
$\epsilon_{\parallel}$ at 20 °C	9.09	5.71
$\Delta\epsilon$ at 20 °C	4.52	1.52
k_{11} at 20 °C$/10^{-12}\,\mathrm{N}$	14.1	16.6
k_{33} at 20 °C$/10^{-12}\,\mathrm{N}$	11.9	13.7
k_{33}/k_{11} at 20 °C	0.84	0.83

This example of the development of liquid crystal mixtures has been recounted at some length to demonstrate the limited but valuable part played in the process by calculation and prediction, and the decisive contribution made by experiment. Gradually knowledge is accumulated of the behaviour of individual components in the system under study, and this in turn leads to mixtures with better properties. The progressive improvement of the low-viscosity broad-range mixture ZLI 1565 at Merck is an excellent example of the same process.

PROCESS RESEARCH AND DEVELOPMENT

Once the decision has been taken to use a particular component in a mixture, it has to be manufactured in the appropriate quantity. When originally synthesized in the laboratory, the route most convenient for the research chemist will have been used. The industrial chemist, on the other hand, seeks the cheapest synthesis, although the options available are limited by several considerations.

1. The raw material must be available from more than one source in case of a breakdown in supplies from a single manufacturer.

2. Solvents or reagents that are dangerously inflammable or toxic must be avoided.

3. The composition of the aqueous effluent that will be discharged after treatment into sewer or river is subject to strict control, and chemicals that are difficult to treat or will interfere in the treatment of others have to be avoided. Similar control must be exercised over gaseous effluent.

4. The chemical steps in the manufacturing process must be accommodated in existing plant if possible to avoid the necessity for capital expenditure on new equipment, at least until present spare capacity is used fully. Extremes of temperature and pressure have to be avoided.

Fortunately, organic chemistry is a sufficiently flexible science to allow these difficulties to be overcome. The development of an alternative synthesis of 4-alkoxy-4'-cyanobiphenyls (figure 5) to the original used by Gray *et al.* (1974), one that avoids the objectionable intermediate 4-nitro-biphenyl, is an example from the history of cyanobiphenyls.

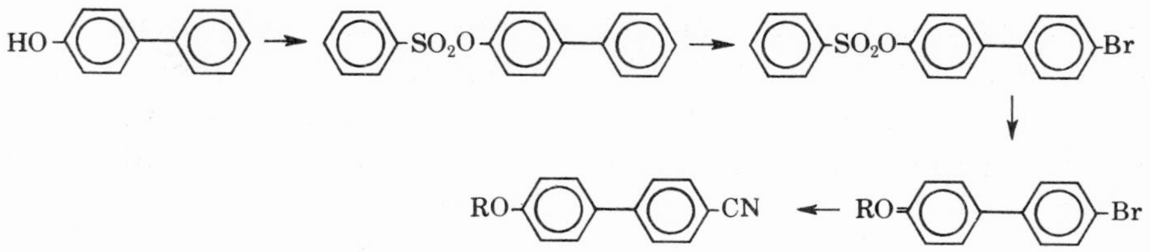

FIGURE 5. Synthesis of 4-alkoxy-4'-cyanobiphenyls used for their production.

Research into methods of manufacture is performed in all the companies supplying liquid crystals, and the scale of the effort may be comparable with that devoted to making mixtures. The discovery of new structures continually poses new problems that have to be solved under stress, because of the desire to commercialize the discovery quickly. The work rarely receives any publicity yet it determines the profitability of the chemical and thus plays a vital part in the sequence by which liquid crystals reach the customer.

PRODUCTION OF LIQUID CRYSTALS

The present annual production of liquid crystals is 6–7 t throughout the world. This quantity is divided among many components, of which 5CB is probably the largest single item, although less than 10 % of the total. Production is carried out batchwise in enamel-lined or stainless steel reactors of the type used generally for the manufacture of organic chemicals. The high purity of the products is achieved by rigorous quality control over the raw materials, by purifying inter-mediates so that impurities are not carried over from one stage to the next, and by lengthy purification of the final liquid crystal, including treatment to remove ions and raise the resistivity. The syntheses often involve many steps, and the total time of processing may be 3–4 months. This makes the organization of production a complicated matter, particularly as many components have to be manufactured simultaneously.

The materials companies and the display manufacturers have collaborated over the past decade to create a new technology, and many of the advances have originated in the chemical side. The display companies get the best of the bargain because their output has more than 10 times the value of the liquid crystal they use, but the materials companies will continue to meet the challenge of providing better liquid crystals as long as markets can be found for their products.

The experimental work was carried out by Mr P. Balkwill, Mr A. Pearson and Dr I. C. Sage. It was carried out under contract to the Ministry of Defence.

References

Gray, G. W., Harrison, K. J., Nash, J. A., Constant, J., Hulme, D. S., Kirton, J. & Raynes, E. P. 1974 In *Proceedings of the Symposium on Ordered Fluids and Liquid Crystals (166th National Meeting of the American Chemical Society)* (ed. R. S. Porter & J. F. Johnson), pp. 617–643. New York: Plenum Press.

Raynes, E. P., Tough, R. J. A. & Davies, K. A. 1980 *Molec. Cryst. liq. Cryst. Lett.* **56**, 63–68.

Phil. Trans. R. Soc. Lond. A **309**, 239 (1983)
Printed in Great Britain

Concluding remarks

BY C. HILSUM, F.R.S.

The past two days have been hectic, heavy with compressed information. I have no intention of burdening you with more technicalities, nor of attempting to summarize the crucial points, for there are too many. Instead, we should stand back from the detail, and try to assess what has been achieved in our field.

We must first accept that the word 'applications' in the title of our meeting really means 'display devices', for today we have few other practical uses of liquid crystals. Nevertheless, liquid crystal displays have an annual world market of £150 million, and this is predicted to grow by 50% per year for the next 5 years. Three years ago rival subtractive technologies, electrophoresis, electrochromism and electroplating were potential competitors, with much activity in research laboratories. Now their limitations have been exposed, and few discuss them. This domination of the field has been accompanied by an extension of our interest to all three types of liquid crystal. The nematic display is well established, and the newer uses of cholesterics and smectics have been discussed by a number of speakers. The wealth of possibilities is enormous, and this presents a danger, because our efforts may become too diffuse.

We should be comforted by the size of the display market, but more impressed with the growth. Most of this audience will realize that the differential has great significance for them. A static industry, though large, calls for little research. A dynamic industry needs new developments, new effects, new materials. Dr Scheffer was concerned about the ability – and willingness – of manufacturers to adopt new developments, such as the dyed phase-change display. Such caution is understandable. A manufacturer who has mastered a device technology with one material will be reluctant to try another. He will be even more resistant to a completely new device structure, unless this can readily lead to an expansion in his market. His profits are not raised if he simply substitutes one device he makes for another. But if the timing is right, new devices will find a place. The history of liquid crystals shows this, because they were ready when the market called for them.

It is not easy, nor perhaps helpful, to compare either academic progress or device invention in different fields, but we should remember that the first integrated circuit did not appear until 15 years after the transistor, and the MOSFET, the most widely used transistor today, was 2 or 3 years later still. It may be that we have not yet invented our l.c. MOSFET equivalent, nor isolated the electro-optic effect on which this magic device will be based. Moreover, our field need not continue to be restricted to displays. We are beginning to see thermochromics in a number of guises, and some ingenious ideas for optical signal processing involve liquid crystal components.

In 1975, at a Conference on Non-emissive Displays, my concluding address chided the participants for their pessimism about the future, and for limiting their horizons to displays for watches and calculators. My final sentence, 'Small men cannot look over high walls,' was never printed, probably because the editor of the Proceedings did not understand my meaning. During the past two days we have heard from tall men, and the obstacles seem much less impressive.

We have an industry that is firmly based, linked closely with distinguished academic research that is equally productive. I am sure you will leave now with a sense of optimism, and I have no doubts that your optimism is fully justified.